THE END OF THE LINE

THE END OF
THE LINE

HOW OVERFISHING IS
CHANGING THE WORLD
AND WHAT WE EAT

CHARLES CLOVER

THE NEW PRESS

NEW YORK
LONDON

Requests for permission to reproduce selections from this book should be mailed to:
Permissions Department, The New Press, 38 Greene Street, New York, NY 10013.

Originally published in Great Britain, in slightly different form,
by Ebury Press, Random House, 2004

Published in the United States by The New Press, New York, 2006
Distributed by W. W. Norton & Company, Inc., New York

Grateful acknowledgment is made to The American Fisheries Society
for permission to reprint P.A. Larkin's "An Epitaph for the Concept
of Maximum Sustained Yield," originally published in
Transactions of the American Fisheries 106 (1977): 1–11.

LIBRARY OF CONGRESS CATALOGING-IN-PUBLICATION DATA

Clover, Charles.
The end of the line : how overfishing is changing
the world and what we eat / Charles Clover.
p. cm.
Includes bibliographical references (p.).
ISBN-13: 978-1-59558-109-9 (hc.)
ISBN-10: 1-59558-109-X (hc.)
1. Fishery resources.   2. Fisheries.   3. Fishery management.   I. Title.
SH327.5.C58 2006

333.95'6137-dc22   2006012058

The New Press was established in 1990 as a not-for-profit alternative to the large,
commercial publishing houses currently dominating the book publishing industry.
The New Press operates in the public interest rather than for private gain,
and is committed to publishing, in innovative ways, works of educational,
cultural, and community value that are often deemed insufficiently profitable.

www.thenewpress.com

Composition by dix!

This book was set in Fournier MT

Printed in the United States of America

2   4   6   8   10   9   7   5   3   1

For Harry and Jack

CONTENTS

THE END OF THE LINE

INTRODUCTION:
THE PRICE OF FISH

Imagine what people would say if a band of hunters strung a mile of net between two immense all-terrain vehicles and dragged it at speed across the plains of Africa. This fantastical assemblage, like something from a *Mad Max* movie, would scoop up everything in its way: predators such as lions and cheetahs, lumbering endangered herbivores such as rhinos and elephants, herds of impala and wildebeest, family groups of warthogs and wild dogs. Pregnant females would be swept up and carried along, with only the smallest juveniles able to wriggle through the mesh. Picture how the net is constructed, with a huge metal roller attached to the leading edge. This rolling beam smashes and flattens obstructions, flushing creatures into the approaching filaments. The effect of dragging a huge iron bar across the savannah is to break off every outcrop and up-root every tree, bush, and flowering plant, stirring columns of birds into the air. Left behind is a strangely bedraggled land-scape resembling a harrowed field. The industrial hunter-gatherers now stop to examine the tangled mess of writhing or dead creatures behind them. There are no markets for about a third of the animals they have caught because they don't taste good, or because they are simply too small or too squashed. This pile of corpses is dumped on the plain to be consumed by scavengers.

This efficient but highly unselective way of killing animals is known as trawling. It is practiced the world over every day, from the Barents Sea in the Arctic to the shores of Antarctica and from the tropical waters of the Indian Ocean and the central Pacific to the temperate waters off Cape Cod. Fishing with nets has been going on for at least ten thousand years—since a time when hunters pursued other humans for food and killed woolly mammoths by driving them off cliffs. Yet because what fishermen do is obscured by distance and the veil of water that covers the Earth, and because fish are cold-blooded rather than cuddly, most people still view what happens at sea differently from what happens on land. We have an outdated image of fishermen as rugged, principled adventurers, not as overseers in a slaughterhouse for wild animals.

Eating fish is fashionable, and seafood is consumed with far less conscience than meat. Even many "vegetarians" see no irony in eating fish. It has become a kind of dietary talisman for Western consumers. Nutritionists tell us that fish is good for us—the best source of low-fat protein and vitamins—and that the omega-3 fatty acids in oily fish aid in optimal brain function, reduce the danger of heart attacks and strokes, and delay the onset of arthritis and osteoporosis. Studies even indicate that consuming fish slows down the aging process and can help us lose weight because a fishy diet switches off our hunger hormone, making us feel satisfied on smaller amounts of more nutritious food. Models, Hollywood actresses, and socialites don't even need to smoke to stay skinny; they can be satisfied on birdlike portions. All they have to do is eat fish.

Unfortunately, our love affair with fish is unsustainable. The evidence for this is before our eyes. We have seen what industrial technology did to the great whales, the hunting of which is now subject to a worldwide, but not total, ban. I believe we are crossing another watershed in public thinking—namely, what industrial techniques, unchecked market forces, and lack of conscience are doing to inhabitants of the sea. On land a watershed was reached in farming when sprays, fertilizers, food additives, and factory-farming techniques used in the raising of crops and animals led to the collapse of farmers' reputations as custodians of the countryside and guardians of the quality of food we eat. The farmers' image is only slowly being rebuilt, amid much suspicion. Fish were once seen as a renewable resource, creatures that would replenish their stocks forever for our benefit. But around the world there is evidence that numerous populations of fish, such as the northern cod, the North Sea skate, the marbled rock cod of Antarctica, and to a great extent the bluefin tuna, have been fished out, like the great whales before them, and are not recovering. Reassurance from official sources on both sides of the Atlantic that the seas are being "managed" scientifically is increasingly muted and, where it is given at all, hard to believe. Enforcement of the rules that are meant to prevail in the oceans has proved wanting almost everywhere. Even in some of the best-governed democracies, experts admit that overfishing is out of control.

Overfishing has, until now, tended to be a peripheral issue on the contemporary environmental agenda, which has focused on damage to the ozone layer, the buildup of greenhouse gases in the atmosphere, the accumulation of toxic and persistent organic substances, and the erosion of terrestrial biodiversity. Perhaps understandably, the public and the news media in

America tend to rank concern about pollution and contamination that affect human health above concern about conserving wild fish stocks. In one sense they are right. We need to be well informed about mercury and PCB contamination in predatory fish such as tuna, swordfish, and shark. Pregnant women and parents should indeed follow freely available U.S. Food and Drug Administration advice on eating fish to reduce exposure to these contaminants and prevent harm to developing nervous systems in fetuses and learning impairment in young children. Yet it is easy to forget that the overfishing of wild fish has a human health dimension. As catches of wild fish decline we are forced in the direction of the intensive farming of fish, with all the attendant problems that this has caused on land—residues of pesticides and veterinary chemicals, the buildup of contaminants (PCBs and heavy metals from concentrating the flesh of smaller wild-caught fish in fish meal), pollution of the seabed, genetic modification, and the potential for creating diseases that can cross species barriers, infecting wild populations and even ourselves. Beneath the near hysteria about mercury and PCBs in large, long-lived, wild, and predatory fish there is a highly questionable proposition—that it should be possible to eat orgiastic amounts of seafood without guilt, whenever one likes, and without taking any responsibility for the health of the biological systems from which that seafood comes. It is time to recognize the selfishness and tunnel vision behind the concern about mercury and to look more broadly at the problems of the oceans.

As we do it becomes clear, as I have suggested, that a perception-changing moment has arrived. It comes with the realization that in a single human lifetime we have inflicted a crisis on the oceans *greater than any yet caused by pollution*. That crisis compares with the destruction of mammoths, bison, and whales, the rape of rain forests, and the pursuit of bush meat. As a method of mass destruction, fishing with modern technology is the most destructive activity on Earth. It is no exaggeration to say that overfishing is changing the world. Just as the deep sea has become the last frontier, its inhabitants a subject of fascination to filmmakers, so some vulnerable creatures of the shallower seas, such as sharks, rays, and seahorses, are already on a slide to extinction. Overfishing, as a direct result of the demand by consumers in the world's wealthier countries, threatens to deprive developing countries of food in order to provide delicacies for the tables of rich countries, and looks set to rob tomorrow's generations of healthy food supplies so that companies can maintain profitability today.

As fish stocks used in traditional diets crash and others are found as substitutes, overfishing is altering our diet. It is even altering evolution: the Atlantic cod has begun to spawn at an earlier age in response to the pressure of fishing. Overfishing has been, and no doubt will be again, a cause of war and international disputes. It is a force in world trade and international relations, and a corrosive agent in domestic politics.

This book argues that, as a result of overfishing, we are nearing the end of the line for fish stocks and whole ecosystems in the world's oceans, and that it is time we arranged things differently. It takes the form of a journey around the world made in several stages and records many conversations about the problems and potential solutions—a number of which are as controversial as the problems. It reveals the extent of what is happening in the oceans in our name while satisfying our appetite for fish, and shows that the true price of fish isn't written on the menu.

1

NAILING THE LIE

Gloucester, Massachusetts, likes to describe itself as America's greatest fishing port. Its claim is inscribed on the most poignant edifice in Gloucester, the fisherman's memorial on the seawall at Cape Ann. A bronze statue of a bearded fisherman in a sou'wester and oilskins, gripping a ship's wheel, stares out over the harbor and the clapboard houses toward the sea. Though you will find that a greater tonnage of fish is landed today in Dutch Harbor, Alaska, and that Gloucester's old rival, New Bedford, Massachusetts, tops the list of U.S. ports for value because of its shellfish landings, it seems rude to quibble with Gloucester's estimation of its own preeminence, for in terms of human endeavor and suffering America's oldest fishing port has history on its side. The fisherman's memorial, installed in 2002, commemorates 5,300 men from the town who died at sea in the pursuit of fish.

The core of the inscription reads: "These courageous men have been known by names other than fishermen. They were father, husband, brother, son. They were known as the finest kind. Their lives and their loss have touched our community in profound ways. We remain strengthened by their character, in-

spired by their courage and proud to call them Gloucestermen."
The story told in Gloucester has an awesome dignity that arises
out of mass human suffering—so many untimely deaths spread
among the population of one not so very large community. The
toll has its equal only in the memorials to the dead of the Civil
War and the First and Second World Wars.

The roll call of the drowned over nearly four centuries is cast
in raised lettering on bronze panels arranged in a semicircle on
the seaward side of the statue. It begins with the Englishmen
who came in 1625 to handline and trap for cod on the submerged
banks that run from here to Newfoundland. Over the next two
centuries, English surnames are joined by ones from Scandi-
navia and Ireland, often via the Canadian maritimes. In the past
two hundred years, many if not most of the fishermen have
been of Sicilian and Portuguese extraction. You will still find
boats in Gloucester whose crews speak Italian and have their
satellite TV set to Italian stations. I met a man of Italian descent
in Gloucester who said his family had been fishing in Gloucester for eight generations. In New Bedford, the capital of whaling in the nineteenth century and now of the sea scallop fishery,
the fishermen are of mainly Portuguese descent.

You can deduce many things from the number of names inscribed on the Gloucester fisherman's memorial. In 2001, the
most recent year recorded on the bronze panels, two fishermen,
Thomas Frontiero and James Sanfilippo, died at sea. In 1927,
the most terrible year in the past century for the fishermen of
Gloucester, the sea took forty-one. You would rightly conclude
that fishing is one of the more dangerous occupations in the
world. You might also conclude that either it has gotten a lot
safer or there are fewer fishermen going to sea. Both happen to
be the case.

You might be tempted to ask why the number of fishermen has dropped, why the fish dock is half empty, and why many of the buildings around it have an air of dereliction that contrasts with the prosperous-looking tourist buildings and marinas of this attractive seaside town. The reason the dock is underused is because there are fewer boats and many times fewer fish landed in Gloucester than once was the case. The story of the overexploitation of New England's fisheries is one you will find the city of Gloucester still reluctant to tell in any of its public pronouncements, for it is the counterpoint to Gloucester's heroic figure of hardiness and heroism. It throws a rather different light on the bronze fisherman facing out to sea.

The square-jawed fisherman on the Gloucester memorial is how the city still likes to think of itself. There is even a new incarnation of him on the big yellow sign of Gorton's of Gloucester, the largest fish-processing firm left in town, on the Harbor Loop. His cutout figure rises above the rectangular Gorton's sign, which tells you that the company was founded in 1849. Yet there is a rich irony here.

Gloucester has an important, perhaps preeminent place in the history of food preserving in the modern era as well as fishing. For it was here in the 1920s that Clarence Birdseye, considered to be the father of the frozen food business, first built a plant after perfecting flash-freezing techniques in his kitchen. His Birds Eye brand of frozen food gained national distribution in 1930. Birdseye fixed the problem that shopkeepers in the Depression could not afford freezers by leasing them one. His business duly became a model for frozen food commerce across the world. Birdseye chose the site for his plant because it was close to large amounts of cheap and highly perishable fish, the value of which he would prolong by freezing. So I rather expected Gorton's of Gloucester to say it got its succulent fillets of white-fleshed fish just across the road at the Gloucester Fish Exchange, the daily fish auction.

There's no mention in Gorton's literature or on its Web site of where its fish comes from, even though both tell you that the company's fillets come grilled, breaded, and beer-battered, with a variety of flavorings. A helpful employee let me in on the secret. The fillets contain Alaskan pollock, caught and processed by American companies in Alaska and then shipped several thousand miles around the continent frozen in containers. Gorton's also sources its pollock from the Russian side of the Bering Sea, which is processed first in China. It imports farmed shrimp from South America and Asia and assorted other species of fish from elsewhere around the world. The reality of present-day Gloucester is that the port has not been able to supply the volume of fish Gorton's requires since the mid-1980s—because of the collapse of New England's fisheries. American consumers now eat more imported fish than fish produced at home.

There is no memorial in Gloucester, as yet, to the bounty of nature that once existed, the profusion of shad, alewives, scup, menhaden, sturgeon, and salmon the founders encountered when they reached the shores of America. Nor is there one to the whales that existed a century ago or the giant halibut, barndoor skate, and much larger shoals of giant bluefin tuna that existed even within living memory. You might think there was a call for some such commemoration, since every visitor who has heard of the contribution the cod made to New England's early prosperity and campaign for independence must wonder what happened. Perhaps it remains a sensitive subject. But the story of the plenty that once existed and its sad decline is one you have to piece together for yourself.

I traveled to Gloucester from England because I had been told by a knowledgeable friend that New England was twenty-five years ahead of Europe, and probably ahead of anywhere else in the world, in reacting responsibly to the collapse of its fisheries. I would defend that observation. But I have to confess that as I began to research what had happened to the abundance that once existed off New England's shores, I was shocked to discover how great that collapse had been—and how fragile and uncertain that recovery remains.

I arrived in Gloucester at 9 p.m. one day in mid-December. The thermometer was down to 11 degrees Fahrenheit and wisps of snow were falling to join several inches on the ground by the time I rolled off Route 128 into Gloucester and found my way to the Cape Ann Marina. I was desperate for a decent meal, having spent two days grimly eating overly generous portions of tasteless food.

The motel restaurant was closed for the winter. The rooms were surprisingly basic given the capital value of the pleasure craft and the bluefin tuna fishing vessels moored or lifted out of the water around it. The night watchman and the desk clerk discussed the options briefly and pronounced that the only place I could get a square meal at that hour was at Franklin's on Main Street.

Franklin's turned out to be the best place in town. It had a smallish facade. Behind there was the sound of a modern jazz duet being sung. It was dark inside, which was off-putting, but after the achingly hip and rather disinterested staff had cleared a table and produced a menu, I realized this place was hot, at least as hot as Gloucester got. On the wine list were some of the finer New Zealand whites and the best California pinot noirs.

On the menu were enticing dishes involving tuna, salmon, and cod. The cod came seared in butter with a hash of chorizo and peppers.

Now, I won't eat cod from the North Sea, for reasons I'll explain later, but this was Gloucester, capital of New England's cod country, and foodie writers are always trying to persuade us to eat food produced locally. So, with only the haziest notion of the relative abundance of the New England cod and with many conversations about the success of conservation over the past decade in my notebooks already, I decided the cod must be all right to eat.

The cod was excellent. The robust taste of seared cod and chorizo, along with the gooseberry acidity of a New Zealand sauvignon blanc, set me right more than I could have imagined. There were many times, however, over the next few days, when I wondered whether I should have ordered cod at all. The fact that cod was expensive didn't necessarily mean anyone was trying to ensure its survival.

6:00 a.m. a day later. The Gloucester Fish Exchange. Outside, the snow was frozen hard. Inside, the auction was in full flow. The auction deals in fresh fish for the restaurant and catering trade. Nineteen buyers in knit hats and baseball caps sat at desks facing the auctioneer, with cups of coffee to keep them alert, bidding visibly and deliberately to get their fish. As I walked in, the auctioneer was selling large cod in small quantities. These were going for a good price, as were the large haddock. The word was that Legal Sea Foods, the successful restaurant chain that now has twenty-six outlets along the East Coast, was buying hard to meet demand over the Christmas and New Year holiday.

Louis Linquata, a fish buyer descended from three previous generations of Sicilian fishermen, told me he gets up at 5 a.m. as a labor of love. He helps out a fish company that supplies the catering trade owned by two friends. He doesn't have to do this; he is the owner of two successful liquor stores. He just likes buying fish.

"In the 1960s," he told me, "you could walk across the harbor on the boats. We were landing millions of pounds of fish. Now it is a tenth of what it used to be then."

Then, Gloucester had eight fish-cutting houses, each cutting half a million pounds of fish a week. Now the volume of groundfish landed in the dock is a fraction of what it was. The species changed, too. Then, it had silver hake (whiting), which is not seen today.

Louis explained that the fish auction house was only nine years old and took the place of the old arrangements under which the fishermen used to sell on the dock to the people they trusted most. Fishermen used to lose 50 cents to $1 a pound by this method of trading, which went on for years. The auction market was set up, he explains, to save the industry. When fishermen were faced with fewer fish, they resolved to get better prices.

Louis remembers his father landing a lot of fish but not making a lot of money. The 1960s were a buyers' market. There was still plenty of cod, haddock, bluefish, and yellowtail flounder. There were times when there was more fish than the market wanted. Prices were low. Now, as chairman of Gloucester's Fisheries Commission, Louis hangs around to make sure the auction means fishermen get the best price there is and there are "no games." With very little fish on the market, prices are good—when twenty people are looking at each fish, the fisherman is bound to get a higher price. "A little bit of fish is making money," said Louis. He's happy. But I wondered if he might also be saying there is an increased incentive to catch the last fish.

Tim Macdonald, a fisherman and former academic who runs a thirty-eight-foot boat out of Gloucester, arranged to have breakfast with me after the fish auction. He told me that he is now catching about 25 percent of the fish he was catching in the mid-1980s. "That's my experience. That's pretty much across the board. Then you could trawl for two hours and get 200 pounds of dabs. Now you have to trawl four hours for 100 pounds."

The day before, I'd spoken to Frank Mirarchi, who has been fishing out of Scituate, forty minutes south of Boston, since 1962. He said: "We caught thousands of pounds of fish in the 1960s. We used to sell little tiny fish smaller than your hand as lobster bait. Nobody was looking at how many fish were killed, just at the landings. The number of fish killed was probably an order of magnitude more.

"No matter what we did we caught fish. We caught fish just outside the harbor. If we went fishing for cod, we caught cod, we also caught whiting and flounder."

Navigation equipment was more primitive then and the nets were made of manila, or hemp, and required lengthy mending, so fishermen only fished in places they knew the nets would not be ripped.

Frank recalled, "My memory is of a tidal wave of fish across the deck when the pocket was opened. This pocket had a very small mesh size. You were looking for marketable fish. The wastage might have been 50 percent, straight into the mouths of the gulls."

Frank remembers that though prices were very low he still paid his way through college on the proceeds, something you could never do today. Even the captain of a high liner couldn't afford to pay for college. The authorities imposed larger and larger mesh sizes from the 1970s on—the larger the mesh, theoretically, the greater the number of smaller fish that escape. The fishermen realized it was not smart to sell small fish as lobster bait, but still the fish got fewer. Fishermen remember the prices improving around 1990—just when scientists began to say stocks were in trouble. Classical economists would say there was an obvious connection between the leap in prices and the beginning of scarcity. It has been a seller's market ever since.

It is harder to track the cycle of boom and bust before the present generation. The signs of dearth and plenty are in the asides and digressions of the histories. Local museums and history books will tell you how the founding fathers first exploited fish that came to them, such the shoals of herringlike shad and alewives that used to run the estuaries in huge numbers. These remain in a depleted state today. Native Americans hunted large fish on the surface, such as swordfish, with harpoons. They also made fairly extensive use of shellfish. The descendants of the Wampanoag tribe still nurture the much-depleted bay scallop around Martha's Vineyard.

The colonial British and their American successors were more rapacious and seem not to have known when to stop. They all but eliminated the Atlantic sturgeon—a long-lived, not very fecund species—from the Hudson River in the eighteenth and early nineteenth centuries. A remnant population of short-nosed sturgeon survives in the Merrimack and some other rivers. The fate of the slower-growing fish of the New World has been overshadowed by the fate of the great whales, in particular the northern right whale, seen playing alongside the *Mayflower* as it lay at anchor in Cape Cod Bay in 1620. A great feat of the imagination is required to picture the oceans before the Basque and New England whalers got to work. Hundreds of thousands of right whales—the name means the right whale to kill, as it tended to float—were taken over the next two centuries until the population became so small it was not worth hunting commercially, around the 1860s. The right whale was supposedly protected by law in 1935, though there are numerous instances of it being killed after that. It is now one of the rarest animals on earth, with a population estimated to be less than 350. The main threats to its existence today are collision with ships and entanglement in fishermen's nets.

Overfishing affects the largest, least plentiful slow-reproducing creatures first—no matter whether they happen to be mammals or fish. Among the nineteenth-century accounts of cod schoonermen and dorymen who caught groundfish—bottom-living fish—in the Gulf of Maine and on Georges Bank, there are records of a major halibut fishery.

The halibut fishery grew up on Georges Bank by the 1830s. Stern figures in Cape Ann oilskin hats posed for their pictures to be taken in front of halibut the size of a man. The number of vessels fishing for halibut out of the port of Gloucester rose to around sixty and for a decade huge catches were landed, up to 15,000 pounds a day. Oversupply was a problem. The Gloucester Fishing Company collapsed within months of its founding in 1848 when it took more halibut than it could sell. After that brief period of abundant catches halibut were still caught when dragging for something else, but the population of large fish was never seen again. You will be lucky to see any halibut on the market today.

Spencer Baird, the zoologist and ornithologist who, as cura-
tor of natural history at the Smithsonian Institution, played a
large part in cataloguing U.S. natural resources in the nineteenth
century, got interested in fish toward the end of his life. Congress sent him in 1871, as the first U.S. fisheries commissioner, to
resolve a loud dispute in Congress between the fishermen in
Massachusetts and those in Rhode Island. Hook-and-line fishermen were complaining that catches of fish had declined in proportion to the proliferation of fish-catching weirs. Baird took
testimony from many witnesses, which he found of limited
value. More factual information came from his own observations over the course of a summer in his temporary laboratory in
Woods Hole. Baird concluded forcefully, in his commission's
report, that the shore fishes had indeed decreased from about
1865, as a result of the combined action of weirs and predatory
bluefish destroying a large percentage of migratory scup before
they had spawned. He proposed weekend closure of the weirs
and sought Massachusetts's and Rhode Island's cooperation.
Though Baird had done his best to carry out his promise to Congress to solve the fisheries dispute, the scientific case for his
opinions was criticized as flimsy by the state of Massachusetts's
commission for inland fisheries, which was in favor of weir fishing. So, not for the last time, scientific nitpicking became the excuse for doing nothing.

How much have the commercial fisheries of New England
declined since the beginning of industrial times? That seems to
be a question that has troubled minds only recently, after the
measurable declines of the past twenty-five years. Andy Rosenberg, who lives in Gloucester and works at the University of
New Hampshire, has found the closest thing to an answer. He
found a nearly complete set of logbooks for the years 1852–59

belonging to a fleet of schooners sailing from the port of Beverly, Massachusetts, and fishing on the Scotian Shelf, just south of Nova Scotia. Some 236 Beverly vessels fished part of the season there, and an additional 90 for another part of the season. The captains noted where they were, what they caught, and the other vessels they sighted.

Rosenberg and his team of researchers found that the catches of this small fleet were prodigious. Some forty-three Beverly schooners alone—among hundreds of other fishing vessels—caught 8,580 tons of cod in 1855. That is 660 tons more cod than the entire Canadian fleet landed there in 1999. Rosenberg estimates that there were 1.39 million tons of adult cod living on the Scotia Bank 150 years ago, compared to only 55,000 tons today. He believes that the cod has declined by 96 percent in the past 150 years.

I was curious about many details of this story, not least how applicable it might be to other commercial fisheries around the world. I called Rosenberg—who happens also to be the former regional administrator (Northeast Region) for the National Oceanic and Atmospheric Administration's National Marine Fisheries Service—to ask whether he had any estimates of the level of decline that could be assumed to apply to other fish species, or to the cod found nearer to the New England coast, on Georges Bank, or in the Gulf of Maine.

He told me he believes the decline is probably of the same order for most of the major commercial species—cod, haddock, herring, mackerel, yellowtail flounder, and winter flounder. In other words, since the mid-nineteenth century, more than 90 percent of the preindustrial population of large, spawning fish has disappeared. That figure would be about right for the decline of large fish in any comparable, heavily exploited part of the North Atlantic, such as the North Sea. Indeed, Rosenberg sent me a paper from Britain's Lowestoft Laboratory with similar estimates for the decline of large fish there. Fishermen and environmentalists alike would agree that the lowest point for New England and its fisheries was the early 1990s. Since the Sustainable Fisheries Act of 1996 and the closure of upward of 8,500 square miles of the offshore banks to fishing, there has been an upturn for many species, notably the sea scallop, which has gone from being overfished to being the subject of the most lucrative fishery in the United States. With far tougher controls on fishing effort and the use of 6½-inch mesh—the widest mesh used for bottom-living fish in the world—there has been a resurgence of haddock and redfish. There has been an increase of a third since 1996 across the nineteen separate fish stocks that make up the groundfish fishery. These are exciting developments that invite study by the rest of the world.

Yet it must be said that the cod, the original mainstay of the fishery, has not rebounded. Nor has the yellowtail flounder, another mainstay species, which is at even lower levels. Scientists released their first assessment of cod stocks in more than three years in the autumn of 2005. It made grim reading. Despite all the conservation measures, the Georges Bank and Gulf of Maine cod had plummeted by 25 percent and 21 percent since 2001, respectively. The Georges Bank cod were at only 10 percent of the level needed for the population to sustain itself against natural fluctuations or disasters, and the Gulf of Maine cod were at only 23 percent of sustainable levels. Yet fishery managers had devised a plan that allows the continued overfishing of cod—as a bycatch when fishing for other species—until 2009.

Priscilla Brooks, director of the Boston-based Conservation Law Foundation's marine conservation program, told me: "We never really ever seem to control the catch. It is very troubling. Now we are being told that we are going to further deplete the cod in order to rebuild it." What the managers are trying to do is to lower the fishing mortality of the cod and give the fishermen something to catch from species that have recovered. The plan is working on Georges Bank—though the cod has yet to recover—but far too many cod are still being caught in the Gulf of Maine. Brooks believes that it will be difficult to pursue other groundfish while avoiding cod. Therein lies the greatest challenge of all.

The uncomfortable truth is that the fishing grounds off New England have been, despite the best in modern science and political endeavor, a fish mine until the mid-1990s. For some species, such as cod, that continue to be caught faster than they can reproduce, they still are. If this is how one of the best-run fisheries in the developed world is managed, it is not fanciful to suggest that eventually we will be left only with plankton.

I became convinced that the bounty of the seas must be shrinking everywhere around 1991. That year I interviewed Len Stainton, a fish merchant with a conscience, who invited me into his office in Peterhead, Scotland's biggest North Sea fishing port. When we were seated in his office, he told me conspiratorially: "They used to land fish as big as a man. Now you are lucky if you get one as big as your hand."

He took me upstairs to the boardroom and showed me pictures of the market in the early 1980s, when there were twice the number of boxes of fish as there were that day. Fishermen in the 1960s, he told me bitterly, used to push fish off the pier if they failed to make what they considered to be the right price. Now other fishermen were paying for their folly. We agreed the decline in catches must be going on worldwide.

The thing that puzzled me, for many years after that moment, was that the decline visible before our eyes on the floor of the Peterhead fish auction was not reflected in the annual figures for catches of wild fish produced by the global authority on fishing statistics, the United Nations Food and Agriculture Organization (FAO). According to the FAO's reports, the tonnage of wild fish caught every year kept going up and up. Perhaps, I thought, there were more fish out there in the far-flung parts of the ocean than people such as Len Stainton and I had imagined.

As it turns out, those fish did not exist. The FAO's reassuring words were unfounded. What we now know—thanks to a piece of exemplary detective work by Reg Watson and Daniel Pauly at the University of British Columbia—was that global catches, which had risen ever since 1950, began to decline at the end of the 1980s. It took twelve years after that moment for this information to become public, even though it was of vital importance to the world's food supply. Perhaps most telling was that it appeared in the pages of *Nature* rather than being published by the FAO.

This is how the official figures were found to be lies. The FAO had reported global fish catches increasing nearly every year, from 44 million tons in 1950 to more than 88 million tons in the early 1990s. The upward trend apparently continued, despite the collapse of Grand Banks cod in 1992. Despite the FAO's own warning that 75 percent of all fisheries were fully exploited or overfished, the total rose inexorably to 104 million tons in the year 2000.

How come? asked Watson and Pauly. They looked at the

productivity of the sea in every region of the world. The reported catches were consistent with the sea's known productivity in every ocean but one: the waters around the world's most
populous nation, the People's Republic of China. Chinese waters were known to be as overfished as anywhere else, but China
went on reporting rising catches and an implausible total of 11
million tons a year. This, Watson and Pauly estimated, was at
least double what was biologically possible. The reason for the
misreporting was that officials in communist China get pro-
moted only if they have increased production. So production
miraculously increased.

In fact, when more realistic estimates of Chinese catches
were used, it became clear that global catches had been in decline since 1988. Watson and Pauly believe the decline is about
770,000 tons a year, which is consistent with what we know
about the drop in individual fish stocks around the world.
That's a lot of fish. An official figure for this decline does not
yet exist, since the FAO, which now draws attention to the "belief" that China's figures are inaccurate, is still trying to persuade the Chinese government to provide accurate records of
landings.

The discovery that the world is running out of fish, as its
human population continues inexorably to increase, would appear to mean that the human race has run into one of the limits
to growth that environmentalists predicted it would hit in the
1970s. There are ways in which our food supply could be stabilized and increased through fish farming. But, as we shall see,
farmed predatory fish such as salmon depend on wild-caught
fish for food. And small wild-caught fish are being fished as unsustainably as larger fish. What Watson and Pauly's discovery
means is that we have crossed a threshold from an era of com-

placency about the sea's wild fish resources to one of concern. As they put it in *Nature:*

> There seems little need for public concern, or intervention by international agencies, if the world's fisheries are keeping pace with people's needs. If, however, as the adjusted figures demonstrate, the catches of world fisheries are in general decline, then there is a clear need to act . . . The present trends of overfishing, wide scale disruption of coastal habitats and the rapid expansion of non-sustainable aquaculture enterprises . . . threaten the world's food security.

You could argue that Watson and Pauly's paper was as important, in its way, to our understanding of the oceans and its wild food resources as Rachel Carson's book *Silent Spring* was for agriculture on land. It was emphatic evidence that things were going wrong in the human food chain on a scale never seen previously. The discovery has yet to be acted on with the urgency it would appear to require, even in the most responsive democracies.

Meanwhile, the United States, which is better than most at managing its fisheries, now passes on the greatest part of its demand for fish to oceans far from its shores, where it can exert only the most trivial influence over how fisheries are managed. In fairness, it has to be said, it is not alone in so doing.

2

FEEDING FRENZY

Tsukiji Market, Tokyo, 4:45 a.m. Top of the food chain. They say that if it swims in the sea, it will end up here in Tsukiji, the biggest fish market in the world. Most central markets have moved out to the city fringes, and Tsukiji has plans to do so, too, but for now Tokyo's fish market remains lodged in the city's heart. This seems apt in a country that eats seafood for breakfast, lunch, and dinner, and out of intricate vending machines between mealtimes. Tsukiji is more than a market; it is a shrine to what one of its wholesalers describes defiantly on its Web site as the "inexhaustible sea." It is a place where a national obsession is indulged and that foreigners visit to be amazed. Part of Tsukiji's fascination is that it is old-fashioned and therefore reveals a lot about Japanese custom. Tourists zealous enough to get up in the dark for their fix of Japanese food culture face multiple threats to life and limb. Motorized three-wheeled trolleys are driven at a frantic pace between auction and stalls. Tractors fitted with giant scoops deposit rock-hard frozen tuna in the back of trucks. Normally polite Japanese give no quarter. The best advice is to get out of the way.

Two large areas in this single-story showroom of the sea

are laid out with tuna. A larger wet-fish area sells virtually everything else from salmon to horse mackerel. Peeping out of the few iced polystyrene boxes that are open for show—in a market obsessed with quality, everything is kept at minimum temperature—are yellowtail, squid of all sizes, anchovies, snapper, barracuda, butterfish, and eye-catching scarlet alfonsinos. On the huge number of small stalls over the covered road from the main auction rooms you will find anything you might want to consume and many things you might wish to avoid. Finger-long eels from China squirm in buckets. Live amberjack and Pacific cod from farms around Japan wait in tanks to be slaughtered. The notorious fugu, or puffer fish, swims listlessly, waiting to kill some unfortunate diners with the poisons in its skin or liver if it is not properly prepared. Unfeasibly large, phallic barnacles and fresh mussels await specialist buyers. A whole stall of pink, cooked octopus is neatly set out, suckers skyward, tentacles tucked underneath, looking like sea urchins without their spines. Here sits a specialty I have yet to see on a plate: a tray of garfish, their long beaks tucked within the coils of their bodies. The buyers delight in variety and seasonality. Just in, and attracting good prices, is Pacific saury, a thin, silvery fish shaped a bit like a miniature barracuda.

Here in the most crowded part of the market, where there is still a danger of being hit by a shopping cyclist, marine delicacies air-freighted from different oceans, involving incalculable emissions of carbon, are packed into every inch of space. Stalls specialize: one has pink prawns from Mexico and black tiger prawns from Vietnam and Malaysia, another scallops, lobsters, and crabs from the north Atlantic, another farmed salmon and fresh salmon eggs from Chile and Norway. These stalls also have a role in the most lucrative business of the market—providing raw fish to a city of 15 million people obsessed with the taste of sashimi, the standard version of Japanese raw fish, and the added design and extravagance of sushi, raw fish in its exquisitely presented haute cuisine form. Top-quality raw fish is everywhere in Tokyo: it just arrives on the side when you order a beer. The stalls are where the giants—the tuna and swordfish—will be dismembered and jointed into more affordable chunks after the early morning frenzy of the auction is over.

From around 4 a.m. the wholesalers arrive to inspect the tuna laid out on pallets by the auctioneers' staff. Nearly a thousand fresh tuna, northern bluefin, southern bluefin, and bigeye, will be sold in the next two hours. About the same number, blast-frozen at sea at -58°F, lie frosted white and gill-less on a floor in an adjoining hall. A mist rises around the buyers' white rubber boots as the tuna warm. Fresh bluefins, the most valuable, are slashed along the lateral line so that the deep red color, the sign of peak freshness, is visible. Each fish is labeled with its weight and surmounted by a tiny sample of the most succulent flesh laid out on a piece of paper for the buyers to taste. This courtesy is largely disdained by the wholesalers, who rely on experience, a flashlight, and a tuna hook to examine each carcass for signs of *yake* or "meat burning"—the first sign of decay, which lowers the price.

The bluefin is the zenith of evolution among fish—an extraordinary creature that can swim at more than 40 miles per hour, accelerate faster than a Porsche, produce millions of eggs, and live for twenty years. The secret of its ability to turn on such dazzling bursts of speed is that it warms its blood through a heat exchanger. What matters to the buyers is taste, and the bluefin—unfortunately for it—happens to be the fish equivalent of Aberdeen Angus raised on Argentine pampas. As a result of the demand for its flesh, the eastern Atlantic bluefin is now listed as an endangered species and estimated to be equivalent to the giant panda in its closeness to extinction. The western Atlantic bluefin stock is in even worse shape and is officially described as critically endangered. That puts it in the same bracket as the black rhino.

It's a big auction today because it is the Friday before a public holiday, explains my guide, a student in his mid-twenties named Hide (pronounced "hee-day"). He is wearing a fleece emblazoned with the badge of the Australian tuna company he is to work for in the new year. The auctioneer rings a bell at 5:30 a.m. precisely. The wholesalers, with cap badges denoting their license to bid, surge into the stands, and the auctioneer starts the sale. Meanwhile, another four auctions by other firms are being held on other areas of the floor. Each auctioneer has his own patter, his own particular tone and vocabulary. To a non-Japanese-speaker the auctioneer we are watching seems to be uttering a particularly frenzied torrent of howls and yelps. Each price is made in less than ten seconds, the buyers holding up fingers to indicate what they are prepared to pay, in an unspoken contract based on trust.

The largest fresh bluefin, and therefore the first to be auctioned, is a relative giant of 510 pounds caught on rod and line off Cape Cod. The New England fishery is conducted for sport, but money is another motive for the skippers, who profit from the catch. The headless carcass on sale in Tokyo was airfreighted from Boston, packed respectfully in plastic sheeting and refrigerated in a huge cardboard sarcophagus filled with ice. The west Atlantic population of bluefins, which migrate up from the Caribbean along the eastern seaboard of the United States each year, has been hideously overexploited in the past, and it is now officially estimated to be about a tenth of the size it was in the 1960s. Independent experts such as Carl Safina say that it is probably much less than that. The United States continues to allow bluefin to be caught by purse seine, harpoon, handline, and long line as well as rod and reel. It regulates its fishery by quotas, seasons, gear restrictions, size limits, and limits per trip. But catches continue to decline. The United States awarded itself a quota of 1,294 tons of bluefin in 2004, but fishermen only managed to catch 971 tons, indicating that this quota was probably far too high. Much ingenuity is expended by the fishing lobby to explain why they just aren't seeing fish where they once did. For all these reasons the west Atlantic population has long been of concern to conservationists. Now there is growing alarm, too, about the larger, interrelated eastern population of bluefin tuna that migrates in a great ellipse into the Atlantic from the furthest corners of the Mediterranean.

A new trend is exerting downward pressure on the market today. To our surprise, the Boston fish makes only $8,260, or $16 per pound. This might seem a lot of money for a fish, but exceptionally large, fresh bluefins have been known to make more than $89,000 apiece. Hide's expectations were higher. Top-quality bluefin tuna usually fetch nearly $20 a pound at Tsukiji. Today, even with strong demand, we see the price of sparkling fresh bluefin slump at under $16 a pound. There is a brief surge to just over $18 a pound for one particularly fresh specimen of southern bluefin from Australia. Hide says tuna farming is what has caused the slide in the market.

Tuna "farming" is the only kind of farming I know in which you reap but you don't sow. Tuna fattening, as it should really be called, started in Australia, where it was found that small, overfished southern bluefins could be netted in shoals and released into cages at sea. No breeding takes place, though Japanese research has shown that to be theoretically possible. It is simply uneconomical compared with rounding up wild fish. The fish are fattened on low-value wild-caught fish until the oil in their belly flesh reaches the optimal level and they reach the right price. Then the tunas are shot in the head and shipped to Japan.

Tuna fattening has caught on fast. In under a decade, it has swept the Mediterranean, where the bluefin tuna has been prized since the ancient Greeks gloried in its slaughter and it was used to feed Roman legions in preparation for battle. Fleets equipped with the latest in fishing technology, such as forward-ranging sonars, catch tuna for the farms. They are guided to the shoals by spotter helicopters and fixed-wing aircraft. The fish don't stand much chance. Once encircled, and before the purse-like net has been drawn completely tight around them, the bluefins are transferred to cages. The cages are then slowly towed from where they are caught to their final destination at less than 2 knots, the slow speed essential to ensure the tuna will not be caught in the nets that form the cage walls and drown. Farms have grown up conveniently positioned around the Mediterranean, in Spain, Malta, Sicily, Cyprus, Libya, and Turkey, so that the bluefins may reach the coastal holding pens with minimal casualties. The trade has generated a vast amount of business, not only with Japan and for the airlines who carry the tuna there, but also for makers of nets and sea cages. Ironically enough, what this means at Tsukiji is that a fish that is overexploited in every ocean where it swims is in oversupply. Too many fish are in the market, too few in the ocean.

In the past five years the practice of tuna farming or fattening has migrated from the Mediterranean to the Canaries and westward to Baja California, where bluefin and yellowfin tuna are now being fattened for export to Japan and for the fusion restaurants of Los Angeles, New York, and Toronto. The numbers of bluefin tuna arriving from Mediterranean and Australian farms can be seen from the number of paper stickers on the carcasses in Tsukiji. "There were 30,000 tons of bluefin tuna from the Mediterranean this year. It was very good catching in the Med, but in two years maybe there could be a supply problem," says Hide.

What Hide describes euphemistically as a supply problem means that the breeding stock of bluefins is being slaughtered at an increasing rate. Before long there could be no adult tuna left to catch.

At sea off Cape Trafalgar, Spain. A battle commences, an unequal contest between man and fish that has changed little in three thousand years. The first sign is the crescent-shaped fin. It breaks the surface of the rectangle of water enclosed by the ten wooden barges that make up the *almadraba*, the trap of anchored nets and floats. A muscular swirl is followed by a splash, which is itself followed by another until the water is white with foam. The adult bluefin tuna, heading for the Straits of Gibraltar as they return from their Atlantic migration to the warm waters of the Mediterranean to spawn, have discovered there is no escape and have begun to thrash around in a frenzy.

Forty fishermen stand and watch as the net is hauled up using hemp ropes and blocks rigged to short masts. Then the waiting figures in orange and blue oilskins spring into action. Seizing hand gaffs and boathooks, they guide, impale, and skillfully haul the giant tunas aboard. The decks of the barges flow red as the wounded tuna flap out their lives in the hold. The air fills with salt spray and flecks of blood.

An adult bluefin tuna can weigh 1,300 pounds, but most of the ones caught today are about 550 pounds—and it takes three men and quite a knack to haul a single one into the barge. Some of the fishermen leap into the net, now shallow enough for them to stand in, to guide the drowning, disoriented tuna to the gaffs. The flesh spoils quickly, so within minutes of the catch, while a few of the tuna's tails are still drumming on the boards of the barges, the ice boat has pulled alongside and men are shoveling a thin layer of flaked ice over their bloody gunmetal and silver bodies.

The ninety-seven tuna caught today in the almadraba are all mature fish. Fishing with traditional, fixed, and selective gear is a sustainable activity. The trap takes two months to set and fishes from mid-May to mid-July. Only fish that swim near the coast get caught. Those fish that are caught are large adults, as the 24,000 feet of net that guide the tuna into the almadraba have 35-inch panels that allow any smaller tuna to escape. Tuna trapping has remained the same in essentials since the Phoenicians strung nets between islands in the straits a thousand years before the birth of Christ. Five and a half miles away, on the windswept sands of Zahara de los Atunes, the seventeenth-century writer Cervantes stayed in the castle, drank wine, ate tuna, flirted with the women, and wrote with affection about the scoundrels who fished in the *almadraba* on the shore. Yes, tuna were once so numerous that the fish could be caught in *al-madrabas* reached from the shore.

In the Mediterranean/Atlantic tradition—for here we are on Spain's Atlantic coast—tuna is almost always cooked, often with vegetables like peppers and tomatoes. In the Hotel Gran

Sol in Zahara they serve *atún aliñado* as tapas, slices of bluefin salted for three months and preserved in oil. If you order tuna salad for lunch, you will get cooked bluefin tuna—*atún rojo,* or red tuna, as it is called locally—not the tropical variety from a can. Yet most tuna caught today, even in the *almadraba,* goes to Japan. In the Tsukiji market a 550-pound fresh bluefin goes for $5,300 to $12,150, a tenth of what it used to, but still more than the $2,650 it is worth in Europe. Meanwhile the market for bluefin in top-quality sushi and sashimi continues to grow: it now extends to Japanese restaurants outside Japan. There is a growing taste for it in China.

The fate facing the bluefin is best illustrated in the *almadrabas'* catches over the past six years. Marta Crespo, the director general of the company that runs them, tells me that the *almadrabas* caught 5,000 bluefin tuna in 1999. The following year they caught 2,000. In 2005 the figure was less than 900. Crespo says that if the European Union does not do something quickly to stop the overfishing, this way of life will disappear. I think she's right.

While the *almadrabas* use traditional methods and hemp nets, their competitors are ruthless, well equipped, and not above cheating. Purse seiners can catch the same number of tuna in a single haul as the *almadrabas* catch all season. The French and Spanish vessels that do the purse seining are guided to the shoals by helicopters and spotter aircraft. These move from Spanish airfields to Oran, Algeria, in the summer to avoid a spotting ban imposed for the sake of the tuna in the season's main month of June. Scientists believe the legal minimum weight for tuna—now 22 pounds—is widely flouted. Certainly, the purse seiners' catch are often so small that they have not yet spawned. The small ones go to supply a ready new market for tuna steaks in France.

Somehow—it is not clear exactly why—many tuna farms operating in the Mediterranean qualify for EU aquaculture subsidies, even though they do not breed tuna and are in direct competition with existing, more sustainable fisheries. Since 1994 Spain has paid out $7.9 million in subsidies to tuna farms. In response to this financial stimulus, and the draw of the Japanese market, Mediterranean tuna farmers caged 23,100 tons of tuna in 2003 compared with virtually none five years before.

The World Wildlife Fund (WWF) says the authority responsible for regulating the fishery, the Madrid-based International Commission for the Conservation of Atlantic Tunas (ICCAT), has been slow to accept that what it once dismissed as a "postharvesting" practice is actually driving the industry. Meanwhile, ICCAT's laissez-faire policy appears to have favored purse seining—the most damaging method—above all other methods of fishing. No wonder ICCAT is jokingly known to conservationists as the International Conspiracy to Catch All the Tunas.

The effect of tuna farming is being felt by other fishermen along the Spanish coast. Carboneras, on the coast of Almería, has a fleet of forty long-line vessels that fish with lines up to 18 miles long for tuna and swordfish. Though far from the most sustainable form of fishing, long lining with large baited hooks is more selective than purse seining. It catches larger fish. The landings are also more easily counted. The long-line fishermen in Carboneras believe the EU should be doing more to control the quantity of fish the purse seiners catch.

Further north again, at Garrucha, Juan Cervantes, the head of Spain's *cofradías de pescadores* (fishermen's associations), has a file three inches thick of correspondence about the tuna farms, which his members oppose for a whole variety of reasons, including the pollution they cause and the introduced diseases that can be shipped in from other oceans in their food fish. He jokes that the tuna are all now in the farms. "I don't know where the French vessels go to catch them. There are none here."

The greatest number of tuna farms anywhere in the Mediterranean is at Cartagena, in Murcia, a couple of hours northeast of Carboneras. You can see the pens below the cliffs covered in wild thyme that were mined first by the Carthaginians and then the Romans for lead, zinc, and silver. There is an *almadraba* here too, now disused. Pedro García, the head of the conservation agency Association of Naturalists for the Southeast, takes us to see the cages in the sea, just over the hill from the La Manga Club, one of the developments built for foreign tourists that are devouring what remains of the Costa Blanca's native coastal forest. In Cartagena harbor I try to talk to the tuna farmers on a supply vessel, without success. Pedro says, "If the bluefin was a mammal, not a fish, people would say, 'Oh my God, we are destroying it, it is becoming extinct,' but because it is a fish they do not care."

Seen from the perspective of another ocean, what happens on the auction floor at Tsukiji is a scandal. Tsukiji may be one of the few fish markets in the world that does not smell of fish— the Japanese obsession with cleanliness extends everywhere, even to the outer alleyways and gutters, which harbor a stench in most markets around the world but here are hosed down and scrubbed several times a day. But Tsukiji stinks all the same, for its daily trade is pushing bluefin tuna daily closer to extinction, just as the market for ivory and rhino horn did to the elephant and the rhinoceros in Africa's grisly epidemic of poaching in the 1980s. The bluefin is a textbook case of unsustainable, underregulated global trade—the sort of thing people tend to mean when they talk about globalization.

Yet there is a perfectly workable global solution. The Atlantic population of bluefin is now listed by the International Union for the Conservation of Nature (IUCN) as an endangered species. And so many conservationists believe that its trade should be regulated under the Convention on International Trade in Endangered Species (CITES). CITES already regulates trade in ivory and caviar—and, just now, has virtually banned trade in both. Moves to include bluefin tuna under CITES were made in 1992. The proposal was made by Sweden, which once had a run of giant tuna off its coast. Sweden's proposal was headed off by self-interested lobbying from the United States and Japan. It can be only a matter of time before an attempt is made again.

In the meantime ICCAT has once again disappointed conservationists by leaving quotas for the east Atlantic stock (the one that spawns in the Mediterranean) at 32,000 tons, when ICCAT's own scientists recommended a maximum of 26,000 tons. ICCAT's population estimates are thought by other conservation bodies to be too high, given that they understate uncertainties over the number of tuna actually transferred to pens and the inadequacy of the catch returns made by countries such as Libya and Turkey. Nor has the so-called conservation body yet acted upon the latest scientific studies that show the Atlantic bluefin population can no longer be separated into distinct breeding stocks for the east Atlantic and the west Atlantic. A study using isotopes found in the tunas' otoliths (ear bones) to distinguish eastern and western bluefin stocks found that an astonishing 43–64 percent of the bluefin tuna collected in the western Atlantic originated from nurseries in the east. Yet

ICCAT continues to "manage" the tuna population on the assumption that there are two distinct stocks, western and eastern, while it seems obvious to any layman that the health of the population in the west depends on a healthy population in the east.

Why, you may ask, does so little happen at ICCAT's headquarters in Madrid when there is such clear urgency? The politics of bluefin tuna is not edifying. ICCAT is mostly a college of vested interests. Of the key players, France has the purse seine vessels and Spain has the farms, as does Italy, where many are controlled by the Mafia. The United States (whose Bill Hogarth, head of the National Marine Fisheries Service, is now ICCAT chairman) has its own East Coast sport fishery. Japan has the market and its own long-line fleet that operates in the mid-Atlantic.

Meanwhile, each country has its own ideas about how to tackle the problem of bluefin overfishing. The United States actually has its own twenty-year recovery plan—but the population estimates it is based on appear overstated when you consider that fishermen can't catch their meager quotas. The EU has tightened up a few things, at glacial pace: farms now have to be licensed, and each member state must have a sampling system to check that the fish in the cages are not too small. Japan, which is waking up to the problem, has said it will not import unless these measures are in place.

The question of setting a sustainable catch quota for the world's most endangered tuna, however, has been postponed at the time of writing. ICCAT has many sad parallels with the International Whaling Commission, a whalers' club that stood by and allowed the blue whale to be hunted to the verge of extinction. The pace of ICCAT's decision making is failing to keep up with the bluefin's decline.

Marta Crespo is not giving up without a fight. Crespo has appealed to Madrid and to the European Union to help the *almadrabas*. She asks why the United States has a plan for restoring its bluefin stock to its former level while the EU simply goes on conniving in unsustainable slaughter. She points out that the EU bears a large share of the blame for setting the new quotas, 19,800 tons of which were awarded to itself. The sight of a fisherman asking for tougher quotas is unusual. Perhaps it is a sign of the times. Marta says the EU has done effectively nothing to protect a form of fishing that has existed in balance with nature since the dawn of European civilization—and which might therefore be considered civilized—from rampant opportunism, unchecked market forces, and unethical practices.

It seems a justifiable accusation. The trend for all the large predatory fishes, the giants of the deep, was set out chillingly in an article in the science journal *Nature*. The authors took ten years to evaluate all the major fisheries in the world and found that only 10 percent of the stocks of large fish—tuna, swordfish, marlin, and large groundfish, such as cod, halibut, skate, and flounder—present in 1950 were left in the ocean. Most worryingly of all, perhaps, the study showed that industrial fisheries take only ten to fifteen years to reduce any new fish community they encounter to one-tenth of what it was before. Ransom Myers, a fisheries biologist based at Dalhousie University in Canada, and Boris Worm, then of the University of Kiel in Germany, used data from the Japanese long-line fleet, which fishes in all oceans except the circumpolar seas. They found that where long liners used to catch ten fish per one hundred hooks, they were now lucky to catch one.

The solution suggested by Myers and Worm was simple but drastic: they called for the number of fish killed every year to be reduced by 50 percent in the populations most at risk. The question of how to make this solution politically acceptable is one of the great problems of our time. One of the ways to protect the sea's big fish is to create a network of sufficiently large areas where no fishing happens at all, to act as reserves.

As Myers points out, this might sound draconian if you were a fisherman, but the alternative is a world in which tuna, sharks, and swordfish are merely memories:

> We are in massive denial and continue to bicker over the last shrinking numbers of survivors, employing satellites and sensors to catch the last fish left. We have to understand how close to extinction some of these populations really are. And we must act now before they reach the point of no return. I want there to be hammerhead sharks and bluefin tuna around when my five-year-old son grows up. If present fishing levels persist, these great fish *will* go the way of the dinosaurs.

As the big fish are fished out, a process of substitution goes on. Big fish are replaced on the menu by smaller fish that are, for a while, more plentiful because the bigger ones—their predators—have been removed. If they can still be obtained, fish from overseas replace big fish from home waters. This process goes on from Tokyo to Seattle and from Gloucester to Madrid. Many scientists have attempted to find a description for this phenomenon of substitution, which continues to provide the impression of plentiful fish supplies. Daniel Pauly calls it "fishing down food webs." In Pauly's view, all we will eventually be left with is jellyfish and plankton. And if you think we will then stop fishing, Pauly has news for you. He reports that one Georgia fisherman is already making a living by sending 50,000 pounds of jellyfish a week to Japan, where the sting is removed and it is turned into a kind of wafer.

I was talking to Donal Manahan, an Irish-born professor of biological sciences at the University of Southern California who was an undergraduate at the University of Dublin in the 1970s. He told me: "I remember this lecturer in my first year saying, 'In your lifetime all the world's fisheries will be gone. They will have collapsed or be in decline.' I had no idea at the time how right he was."

While scientists predict a grim future for the world's remaining wild fish stocks, the consumers of the United States, the EU, and Japan all show themselves ready to pay more to eat them. A sure sign of diminishing supply is that the price keeps going up. Fish has become more expensive in real terms as demand has increased. The rise in the price of seafood over the past thirty years is even more remarkable when you consider that the prices of chicken, beef, pork, and dairy products have fallen dramatically as a result of technological advances in farming.

The only exception is when there is a temporary glut because the wild stock is being wiped out faster than people can eat it. It was like that once with the halibut off New England, as it is now with the farmed bluefin of the Mediterranean.

It is a Western myth that bluefin tuna is impossibly expensive in Japan and eaten only by the very rich. If it were, the trade might actually be easier to control. In fact, most bluefins will be consumed by hundreds, even thousands, of people in tiny amounts. The outer perimeter of Tsukiji is circled by restaurants, which Hide and I reached at 9:00 a.m., longing for breakfast after being up since the early hours. We fell into the first restaurant we came to. Hide chose cod and cabbage soup, which sells at about $5.25. The menu also listed a bluefin sashimi "set" for $9.50. I ordered it on a dual impulse of hunger and curiosity. The pile of dark red flesh with a sensuous texture and a rich taste, weighing in total around 5 ounces, came with an elegant garnish of twirled spring onion and cucumber, as well as some wasabi and soy sauce. On the side were a box of rice and a bowl of miso soup. The taste was delicious and has remained memorable, as has the guilt that struck me once I had eaten it. That, I resolved, was enough bluefin for a lifetime, or until bluefin numbers miraculously increase. If that is to happen, the Japanese—and everyone else who likes Japanese food—are going to have to get used to eating fewer fish.

3

ROBBING THE POOR
TO FEED THE RICH

Dakar, Senegal, West Africa. The *Vidal Bocanegra Cuarto,* a 149-ton freezer trawler registered in Huelva, Spain, rode high on the black, oil-polluted water of Dakar harbor like a swan. The Spanish vessel had been painted from bow to stern in a light gray, probably on its most recent voyage, as paint cans were still lying on the deck. There was not a scrap of rust anywhere, except on the steel of its trawl doors and hawsers, which were constantly exposed to the sea. It stood out from the working Senegalese trawlers, with rivulets of rust across their paintwork, and from the many wrecks hitched together like pontoons for repair or scrap. Most corroded of these, and strangely sinister in the evening light, was the *North Sea I,* a listing, distant-water trawler dating from the era of the last cod war between Britain and Iceland in the mid-1970s. It was apt that this old buccaneer should end its days here.

As the press delegation landed to examine the Spanish trawler more closely, its skipper appeared and invited us aboard. Pepe José Vidal Acuña was rust-brown himself from years at sea. He had put in to harbor for supplies and to land his catch—large shrimp, crab, and lobster to be air-freighted that day to Spain's discerning fish markets. Mediterranean countries pay the best prices for shrimp, a taste their home waters formed but, ironically, can no longer supply. Happy to be off duty, Pepe invited us to explore his vessel. Bo Hansson, a Swedish fisherman and journalist, pointed admiringly at the array of screens on the bridge. The craft boasted two of everything: two fish finders, two satellite navigation systems, two sonars, two computers, all to avoid a premature return to home port. The *Vidal Bocanegra Cuarto* lacked for nothing—she was a well-run, immaculate fish-killing machine.

"The Spanish fish all over the world," Pepe declared with an expansive gesture, encompassing Angola, Mauritania, and Morocco, among others. In Senegal his vessel fished with a seventeen-member Senegalese crew and three Spaniards, who were clearly in charge. Pepe spent six months of each year catching shellfish off the coast of West Africa, and he caught enough on this long expedition to enable him to spend the next six months at home.

The day before in these same waters our motorboat had pulled alongside a large *pirogue*, a traditional Senegalese seagoing canoe. The crew of eleven fishermen chanted songs as they hauled in their purse seine. It came in empty, except for a tiny broken seahorse and a few sea urchins. One of the young crew, Lamin Saar, showed us a scattering of sardinellas in the hold: "Today is very bad. We won't get anything except the price of the fuel." He said that when his grandfather was fishing, *every* day was a good day. On a fortunate day now, each crew member can earn as much as $18, but this happens no more than once a month. The next day at M'bour, a large fishing port down the coast, a blue-robed elder in gold-rimmed dark glasses came to meet the foreign press. He wanted to tell us his opinion of the deals signed by his government, which allow trawlers from the EU, Japan, and Taiwan to fish in Senegal's waters. "Poverty came to Senegal with these fishing agreements," he said succinctly. Scientists show that there is more to this view than nostalgia.

Fed by one of the Atlantic's great upwellings, the waters of West Africa are among the world's richest, with more than 1,200 species of fish. Upward currents, created by the force of trade winds rushing off the desert of Mauritania and out to sea, draw nutrients to the surface from the deep. The nutrients stimulate plankton, the base of the whole marine food chain. Senegal's fish markets, usually to be found on a shore strewn with brightly painted *pirogues,* provide evidence of this extraordinary ecosystem. Barracudas, 5 feet long and prized for their firm flesh, lie piled on tables. There are also pink African sole, sea bream, *thiof* or white grouper, and *capitaine,* called giant African threadfin in English. There are strange, slabfaced fish and jelly-fleshed eels, which have no English name. Mothers hold up juvenile, plate-sized groupers, so beloved of the restaurant trade, that should not have been caught, as they will not have spawned. Smiling children offer a few fish on a plate or a single octopus. There is something for everyone. The poor in Africa eat small fish, sardinellas and other pelagic (shallow-swimming) fish. These are caught by fishermen in *pirogues* with handheld purse seines and dried on raised tables by the women before being sold inland.

For about a decade scientists have been warning that the fish stocks of West Africa's continental shelf are overexploited and that some, such as grouper and sea bream, are actively facing collapse. There is absolutely nothing in place to prevent their fate from resembling that of the once-innumerable northern cod off Newfoundland. Hake, too, caught along with deep-sea shrimp further down on the slopes of the continental shelf, is overfished. If stocks collapse and local people starve, Europe—including the former Soviet Union, which fished heavily in these waters until the end of the 1980s—will bear the largest share of the blame.

The European Union has had little success looking after fish in its own waters, but it has strong fishing traditions, and fishermen vote and know how to lobby. So the EU spends $227 million a year buying access for European fishermen to distant waters all the way from the Arctic Circle to the Falkland Islands. It recently signed a new deal—dubiously known now as a partnership agreement—with several West African countries, including Senegal. The beneficiaries of these little-publicized agreements are distant-water trawlers from Spain, France, Italy, and Greece.

Among the countries the EU has signed deals with is Angola, where millions of people risk starvation. This is a matter of indifference to the country's elite, who earn hundreds of millions of dollars a year in oil revenues. They also receive $32 million from the EU for allowing eighty-five of its vessels to fish for tuna, shrimp, and demersal (bottom-dwelling) fish, which, from the EU's point of view, sounds like an extraordinary bargain. It is unclear whether British or German tourists on their package vacations in Spain, or indeed the Spanish themselves, are aware that the shrimp in their paella are taken not from the Mediterranean but from seas belonging theoretically to some of the starving in Africa.

Fishing in the waters of countries that are distracted by social unrest is a favorite trick of the EU fleet. The Spanish and French tuna fleets regularly fish in the waters of Somalia, the only country on Earth that doesn't currently have a government. The European Commission recently renewed its fisheries agreement with Ivory Coast when that nation was in the throes of civil war.

There is more than a faint imperial echo in fishery agreements, because they are usually entered into by unequal parties. Some agreements are better than others. The United States has fewer than the Europeans, but it does pay for its tuna fleet to fish in the South Pacific under a multilateral agreement concluded with the Forum Fisheries Agency, which represents a patchwork of nations large and small, from Australia and New Zealand to the Solomon Islands and Micronesia. Historically, cooperation over tuna was seen by the United States as a way to strengthen ties with south Pacific countries. The skipjack portion of the stock is currently reckoned to be in good shape, and independent reviewers have pronounced the U.S. treaty fairer than most because it has put more money into the pockets of the small Pacific nations—more than $200 million since the signing. Attempts to persuade Japan, Taiwan, and Korea to enter into a similar treaty, instead of concluding their own bilateral agreements with small Pacific nations, have been unsuccessful. However, it is highly doubtful whether the price paid by the United States represents a fair "rent" for some very valuable fish. And a significant portion of the funds paid by the United States come from budget of the U.S. Agency for International Development (USAID), so it is more accurately described as a subsidy. Such subsidies are supposed to be phased out under world trade rules.

Distant-water fishing nations rise and fall. The huge and subsidized Soviet fleet plundered its way as far as Antarctica, then collapsed when the Soviet Union did. Russian capital is behind many trawlers that fish illegally today. There was a time after the Second World War when Britain also maintained a large distant-water fleet, sailing out of Hull and Grimsby, towns that then vied for the title of greatest fishing port in the world. Then British trawlers fished for cod all the way to the Arctic Circle, in what became the territorial waters of Iceland, Norway, and Russia. Now all but a tiny fragment of that industry is gone. The free-for-all ended with the 1982 signing of the UN Convention on the Law of the Sea (UNCLOS), which gave a legal basis to attempts by Norway and Iceland to close their waters to foreign fleets to conserve stocks—which in Iceland's case sparked three "cod wars" with Britain. The Nordic peoples won. Iceland and Norway were too rich and too democratic to give away their birthright cheaply. African countries—or rather their ruling elites—have seen their waters more as a ready source of instant foreign exchange. So the neocolonial days live on for Spain, which maintains a fleet of more than two hundred trawlers off the coast of West Africa, largely at other EU nations' expense. The European Commission argues that fair access agreements are encouraged under UNCLOS, which established all countries' exclusive right to resources within their 200-mile limits.

The theory is that fisheries access agreements established a means for poor nations to profit from the harvesting of a surplus they did not have the technical means to harvest themselves. Such a surplus may theoretically still exist in the waters off Mauritania, a country with a relatively small artisanal fleet equipped with traditional vessels and a vast continental shelf, but not in the waters off Senegal. Senegal has an industrial fleet of its own, catching 528,800 tons of fish and shellfish a year, and a vast artisanal fleet, based on *pirogues,* which catches more than 352,000 tons. Fishing is a source of employment for 600,000 people, a significant proportion of the coastal population. This population is increasing as a decline in rainfall since the 1970s has brought more people from the interior of the country. As Calixte Ndiaye, a college professor who runs a small, local organization that works with fishermen, told us, "The sons of fishermen were taught to swim by the time they were two. Now you see many fishermen who can't swim."

There is no longer a surplus of fish here. Daniel Pauly, the University of British Columbia professor who proved global catches were falling, has spent years studying tropical fisheries. He has calculated that fish stocks off the west of Africa have declined by 50 percent since 1945, when industrial exploitation effectively began. There is increasing conflict between industrial and traditional fishermen. The access agreements specify that only artisanal fishermen may fish within 6 miles of the coast. But these rules are routinely broken. Industrial trawlers, both foreign and Senegalese, are known to venture closer to the shore than allowed at night. The *pirogue* fishermen, equipped with modern nets and outboards, like to venture further out to sea. They concede that generally the Europeans tend to give them a wide berth if they are seen, but the Taiwanese often ignore torches lit by the seagoing *pirogues*. There are many fatal accidents when crews fail to notice each other—in 1997 nine fishermen were killed in a collision between an EU trawler and a *pirogue*, bought for poor villagers by a French development charity.

The argument of the EU and their partners to the Senegalese government is that the industrial trawlers and traditional fishermen are catching different stocks. It's not true. Foreign trawlers catch bottom-dwelling fish on the narrow continental shelf in up to 660 feet of water, the very same fish that traditional long-liners in their pirogues catch more selectively. It is true that foreign vessels in Senegalese waters do not catch the shallow-swimming fish, such as sardinellas, that the *pirogues* catch. These fish, however, migrate northward to the waters off Mauritania, where they are caught by a fleet of international factory ships. Monitoring in the Senegalese waters is rudimentary—there is one (frequently grounded) fisheries inspection plane—and there is plenty of scope to cheat the system.

The most telling point about the EU agreement with Senegal is that it imposes no catch quotas to conserve stocks. Instead it sets the total tonnage of vessels that may fish in Senegalese waters at any one time. The 165–275-ton trawlers in Dakar harbor can catch all they want provided they use the right mesh—smaller and less selective than in equivalent fisheries in the EU. The European fleet declares catches of 13,200 tons a year, but this is widely disbelieved. Reports by WWF estimate that the EU's catches had to be nearer 88,000–110,000 tons a year, up to eight times the declared total.

That is just the weight of the fish *landed*, not the ones that go over the side. Fishing for shrimp also nets a number of fish as bycatch. Fishermen who trawl shrimp for Spanish paellas admit that they account for around 15 percent of their catch. The other 85 percent is fish, some saleable locally, much not saleable at all. The *Vidal Bocanegra Cuarto* fishes with a 1½-inch mesh net and might expect to catch 44 pounds of shrimp—big creatures up to 8 inches long—and 110 pounds of saleable fish per hour. The boat might expect to catch the same amount of undersized shrimp, juvenile fish, and inedible species, such as sea urchins, undersized octopuses, and crustacea, which would be dumped over the side. Rather disturbingly known as "trash fish" by the industry, they include juveniles of all the most locally important fish species, particularly bream and grouper, the stocks in the worst trouble. So industrial trawlers such as this one can justifiably be accused of stealing Senegal's future. The shrimp they fish for are declining, as are the trash fish they catch in the same nets.

Jacques Marec, the French-born head of a fishing fleet of nineteen trawlers based in Dakar under the Senegalese flag, observes that catches of shrimp are declining overall by 330 tons a year; in 1983 each trawler was catching 165 tons of shrimp, but now it is catching 44 tons. This has occurred partly because of increased competition, but mostly because of the absence of fish. Senegal's industrial fleet clearly shares as much blame for overfishing as anyone else, but Europeans cannot offload their responsibility that easily. The home fleet fishes almost exclusively for export to the European market. There's no escaping the fact that destruction of West African fish stocks has arisen mainly from demand in Europe, followed closely by Japan and Taiwan—a fact of which the consumers in those countries are almost entirely unaware.

The European Commission likes to proclaim that access agreements can be a way of helping the African countries involved. But it continues to negotiate them furtively, like a guilty secret. Its latest "partnership" agreement with Senegal was conveniently negotiated just before the World Summit on Sustainable Development in Johannesburg in 2002, where the EU undertook to implement recovery plans for endangered stocks by 2015. The Senegalese deal was brokered, with scant publicity, six months before ministers agreed to improve the terms of such agreements in future. Dr. Ndiaga Gueye, director of marine fisheries for the Senegalese government, has no illusions about what he was negotiating. "We are talking about commercial agreements," he said. He confirmed that in the course of an eighteen-month negotiation, which he headed, the EU actively resisted numerous conservation measures and drove a hard bargain on price. He declared that under the terms of the agreement, Senegal would earn far less if it were to impose quotas to save its endangered stocks. The Japanese had provided Senegal's single research vessel, while the EU had provided no help whatsoever in managing its stocks.

So why did Senegal sign? Gueye indicated that the money, $75 million, was a tidy sum to a poor country. Senegal needed hospitals and schools so that the next generation might learn to be something other than fishermen. He claimed that Senegal was open and transparent about how these revenues were spent, but it still didn't seem to add up. So I asked what else was in it for them. Gueye spoke obscurely of "political and diplomatic reasons." Did EU negotiators make a link between fisheries and aid? That is the clear impression he gave. We also learned that diplomats from the Spanish embassy in Dakar—interestingly, not EU officials—spent a lot of time at the fisheries department before the deal was finally agreed. It is a common assumption in Senegal that there is corruption somewhere in the system.

The local fishermen do like some things in the new agreement. A proportion of all crews must be Senegalese and, instead of being shipped straight out as before, 18,700 tons of tuna a year—caught off Senegal, Gambia, and Mauritania—must be landed in Senegal for processing. Tuna is the one species that the local fleet does not have the means or the experience to catch. All the same, Senegalese trawlermen question the legality of the access agreement now that stocks are clearly overfished, and the artisanal fishermen are angry that the agreement was signed behind their backs. Environmentalists say that cash-for-access deals signed by the EU still stink but are better than previous ones. Julian Scola of WWF says they raise "serious questions about whether these agreements are compatible with the EU's rhetoric on sustainable development and with its development policies which are about eradicating poverty." Environmentalists are caught. They cannot oppose access agreements altogether because if the EU pulled out, it would open the door to murky deals with individual companies or other fishing nations who might behave no better. Instead, they want the terms rewritten to promote conservation.

What makes conservation even more difficult is the growing number of EU trawlers transferring to the Senegalese flag. Local boats face less stringent rules, carrying no official Senegalese observers, as larger EU vessels are obliged to do. In theory, joint ventures between local companies and foreign fleets have to be 51 percent Senegalese-owned. But no one really complies with this rule. "There are people who one day have problems buying long trousers and the next day are the owners of four trawlers," one Senegalese official remarked tartly. All the observers I met had been offered bribes by EU skippers to turn a blind eye to the rules. The EU has said it wants the transfer of EU vessels to foreign flags to stop, but characteristically it has given them a year's grace period, and when the new rules are introduced, they will not be retroactive. Needless to say, joint ventures are proliferating rapidly.

But in this industry a new wind of change is sweeping through Africa, and two fisheries access agreements—at last recognized as the neocolonial practices they were—have been annulled. Namibia was the first to crack down on EU trawlers in 1992. Even after Namibia became independent in the 1970s, Spanish trawlers continued to harvest huge quantities of the country's hake stocks. By then the hake was almost all classed as juvenile (below spawning size), and there was an acute danger of stock collapse. The Namibians hired a helicopter, and local fishery inspectors alighted on the decks of the illegal vessels to carry out arrests. The Spanish were indignant, but gradually the Namibians took control and used their marine resources for their own benefit. Fisheries have since become that country's main engine of economic growth.

Ever resourceful, though, the EU persists in signing up new fisheries agreements with countries that may not know what they are letting themselves in for. It recently signed its first agreement in the Pacific with the Solomon Islands. Somebody in the Spanish or French distant-water tuna fleets, which have the only vessels capable of dashing to the western Pacific on a whim, must have been reading the latest assessment of the world's fish stocks by the UN Food and Agriculture Organization. This says that the western Pacific is one of the few places on Earth where fish stocks are underexploited. Tuna catches are still healthy, with around 1.9 million tons being caught each year. Countries in the region recognized the need to manage growing fishing pressure in 2000 when they signed the Western and Central Pacific Tuna Convention, establishing a commission to limit catches to a sustainable level. It remains to be seen how successful it is now that all the world's main tuna fleets— belonging to Japan, Taiwan, Korea, the United States, the EU, and now China—are fishing in the south Pacific.

Senegal is already far closer to disaster. In many ways a model democracy by African standards, its politicians have some unenviable decisions ahead. Senegal's government recognizes that its stocks are endangered and that something has to be done, but this has yet to be matched with urgency or action. The consensus is that Senegal has five years before its fisheries will collapse. Even if it ejects foreign trawlers from its waters, it faces challenges restraining the growth of its own industrial fleet, a growing proportion of which formerly flew European flags, and its artisanal fishing. The future doesn't look good: globally there are few successful models for managing such a complicated mixed fishery—where many species, both fast- and slow-reproducing, coexist. There are many other African nations in the same position as Senegal—starting with its neighbors in West Africa, but including countries in the Indian Ocean—and countries all over the world who rely on their 200-mile limits as a major source of income. Through model agreements and more informed buying habits, European voters and consumers could exercise important pressure for the conservation of fish stocks, but the European Commission listens only to vested interests and talks of good governance while bribing Africans to persist with unsustainable practices and to allow the pillaging of their waters by EU vessels. The present access agreement with Europe has only hastened the onset of disaster.

Here ends this book's headlong plunge into the disturbing things going on in the world's oceans. You might find that your head is spinning, that there is too much to take in, more than you can possibly worry about as a consumer or voter. I confess that there are times when I feel that way as well. When this happens I try to reflect upon a single sea—the muddy, industrial North Sea that begins 3 miles from my door. It is one of the most heavily used seas in the world. It supports a major oil and gas industry; it was once both the fish basket and main highway for northern European civilization; it is also, though we often forget this, an ecosystem full of wild creatures, and one that has changed considerably. When I begin to reflect on what has been lost from the North Sea I become resolute. I feel the anger rising again.

SEA OF TROUBLES

The produce of the sea around our coasts bears a far higher proportion to that of the land than is generally imagined. The most frequented fishing grounds are much more prolific of food than the same extent of the richest land. Once in a year, an acre of good land, carefully tilled, produces a ton of corn, or two or three hundredweight of meat or cheese. The same area at the bottom of the sea in the best fishing grounds yields a greater weight of food to the persevering fisherman every week of the year.
— Professor Thomas Henry Huxley, address to the
International Fisheries Exhibition, London, 1883

Ten thousand years ago, one of the most productive seas ever for fish—the North Sea—was tundra. The last ice age suspended so much water above ground as glaciers and snow that sea levels around the world were about 165 feet lower than they are today. The present-day southern North Sea was part of continental Europe and settled by humans. Tribes from the area of the modern Low Countries used to cross what is now the sea and follow a funeral route along the Ridgeway to sacred places in southwest England. Sites of Stone Age coastal settlements have been found on what is now the seabed, their middens (trash heaps) containing shells and the bones of fish and sea mammals. The archaeological record shows that early Europeans clung to the shore because the sea was their principal source of food. A diet rich in marine mammal oil would have helped to keep them warm.

The North Sea is therefore a shallow sea, two-thirds of it under 165 feet deep—little more than the height of the Statue of Liberty. One of the first things any visitor notices about the southern North Sea is that it is muddy. The waves are brown-gray in an easterly gale, and blue-green with suspended sediments and plankton in high summer. Visibility for divers is seldom more than a few yards. This is because of plankton growth and because sediments carried down by the rivers and the sand and gravel of the bottom are churned up by the waves and tides. Locally the water can be even muddier as a result of dredging to keep ports clear and to extract sand and gravel. The water clears briefly for about three weeks in May, but in the southern part of the sea, that's about it. The northern part of the North Sea is deeper, dropping to more than 2,300 feet in the Norwegian Trench. In the northern part of the sea, divers will tell you, the water is clearer and the bottom harder—there is what scientists call a hard substrate, which supports a greater range of plants and plantlike animals, together with a greater variety of shellfish. But was the southern North Sea always muddy? In all the time I have spent vacationing on its shingle or sandy shores, swimming in it, sailing across it, or watching ferries, rig tenders, and fishing boats in the evening light, it never occurred to me that there might have been a time when the sea was not cloudy. Only recently have I discovered that the muddiness of the water may well be explained by overfishing.

Vast natural oyster beds covered large parts of the east coast of England at the time of Christ, and oysters were traded throughout Europe even before the Romans invaded Britain. Remnants of these oyster beds remain productive in Essex and Kent, but at a much reduced rate owing to centuries of fishing and a parasite that devastated the population in the twentieth century. What is not often remembered is that there were also vast oyster beds in shallow areas of the open sea, producing a hundred times more oysters than today only a century ago. Nineteenth-century maps show oyster beds 120 miles in length on the German and Dutch side, but the last of these were fished out before the Second World War. Since then there have been no oysters left to form a hard substrate across the bottom. Maybe there have been changes in the Atlantic water entering the North Sea that have made the seabed a less comfortable habitat for oysters. But the simplest and most obvious explanation for the disappearance of the giant North Sea oyster beds is overfishing.

With a covering of vastly more oysters, many of them very large, and a bed of the shells of former oyster generations, the bottom of the North Sea would have been less mobile and its plankton more in demand as food. The bottom would have settled into a hard substrate and the plankton and suspended sediments in the water would have been stabilized by the bivalves as food and glutinous excrement. The hard bottom would have supported lobsters, today less plentiful in those parts of the sea. It therefore takes only a short step to conclude that the water was probably clearer, certainly when one adds to the filtering power of the oysters the larger mussel beds that would then have existed around the North Sea coasts, and the greater areas of salt marsh and shallow estuary that would have caught more of the sediments coming down the rivers. If we assume the southern North Sea was once clear, it follows that increased sunlight would have reached the sea bed, improving the productivity of the ecosystem generally, but in particular the bottom flora. It is probable that the vast areas of mostly featureless sand and gravel at the bottom of the North Sea represent a devastated ecosystem, its natural purifying capacity lost.

Environmentalists from North Sea countries are fond of complaining about eutrophication—the overgrowth of plants and algae, some of which are toxic, because of excessive nutrients dissolved in the water. While it's true that high levels of nitrogen and phosphate come from farm runoff and sewage plants, the influence of overfishing oysters is never mentioned—probably because the groups in question don't know it happened. They, like the rest of us, tend to assume that the sea they experienced in childhood was "natural" and as it should be. But studies of Chesapeake Bay on the other side of the Atlantic, now a eutrophic soup of plankton, show that it was once clear. Oysters the size of dinner plates used to eat enormous amounts of plankton, filtering the water of the bay once every three days. In the 1700s, the oyster reefs were so large that they were a hazard to navigation. Needless to say, these oysters were fished out. The sedimentary record is likely to show that parts of the North Sea had the same filtering capacity. Now it is unlikely that oysters would reestablish themselves, even if they tried. Oysters need to be undisturbed for four years to reproduce, and most suitable parts of the southern North Sea are trawled at least once a year.

Also missing from the sea, though present historically, are a number of large mammals. Largest of all was the gray whale, which feeds on mussels on the seabed and in estuaries. Mussels and oysters would have lived in huge profusion in the delta of the Rhine and in the Wash, especially before land there was "reclaimed" from the sea. Gray whale bones have been discovered around the North Sea, the most recent dating from the early seventeenth century. There were right whales at least until the late Middle Ages. It is fair to assume that there were more dolphins, common and bottlenose, and harbor porpoises but fewer seals than there are today. The laboratory director of the Royal Netherlands Institute for Sea Research on Texel Island, off the northwest coast of Holland, recorded seeing six harbor porpoises at the same time from his office window in the 1950s. Now there are hardly any harbor porpoises in the southern North Sea. I once became dreadfully seasick while on an acoustic survey intended to find them, which concluded that there were none to be found. Records of sightings from ferries to and from Texel Island document bottlenose dolphins that are not there today.

So much has changed, even in a single lifetime. It becomes an awesome task to document what has happened over millennia, both naturally and as a result of human intervention. Traces of even more distant times are present in the North Sea. Trawlers still pull up bones of mammals from the Pleistocene era— bison, musk ox, woolly mammoth, and woolly rhino—and a trade exists in sorting these for museums and private collections. Within our own climatic era (the past ten thousand years), "haul-ups" of walrus by nomadic people would certainly have happened, and may be assumed to be the reason we no longer have that kind of sea mammal. We know porpoises were hunted by Stone Age people who lived and beached their boats on a shoreline that is now several kilometers off the Danish archipelago. Sailing boats, harpoons, traps, and simple trawls may on their own have eradicated some slow-moving species of fish and mammals, just as other oceans lost their turtles and sea cows.

Inevitably, it is easier to list what has disappeared in the past two centuries. When it comes to fish, most of the absentees are those that migrated from salt water to fresh water to spawn, thus multiplying the risks to their survival. The sturgeon is one such migrant. Sturgeon are living fossils, much more common 60 million years ago than they are now. Atlantic sturgeons were caught in the Netherlands until the 1870s, when inventive fishermen employed steam engines to drag nets along the riverbed. The sturgeon of the Rhine delta are now gone. The houting, a migratory, codlike fish, is another absentee from the western side of the North Sea, found occasionally on the Continent. The Atlantic salmon, now in decline, must once have been as plentiful in rivers on the east and west coasts of the North Sea as it is on the Kola peninsula in Russia today, roughly as plentiful as the Pacific salmon species are in Alaska. There it's still said that you can almost walk across the water on their backs. At one time that must have been true of the Thames and Rhine.

Perhaps the biggest fish absentee, the bluefin tuna, disappeared as recently as the 1950s. Tunny fish, as they were called, used to follow the herring shoals. Specimens of stupendous size were caught on rod and line within 25 miles of Scarborough from 1929 until 1954. Then, for reasons that have never been satisfactorily explained, the run disappeared. The pioneer of the rod-and-line sport was one L. Mitchell-Henry, who dedicated himself to catching a giant bluefin after a large fish was harpooned off Scarborough in 1929. He eventually landed the British record, a fish of 851 pounds, from a rowboat in 1933. The sport was such a draw for Scarborough's tourist trade that the council made a clubhouse freely available to the British

Tunny Club, formed in that year. Mitchell-Henry fell out with the club when the sport was overwhelmed by rich socialites, who used large boats and what he considered to be less-than-sporting techniques. The end of a priviledged era came with the Second World War, but not before Captain C.H. Frisby had set a world record for the greatest weight caught in a day, with five tunny weighing a total of 1.25 tons in 1938.

British people had yet to acquire a taste for raw fish, or even tuna steaks, in the 1930s, so most tunny were sold for fish meal. Wasteful as this sounds, the sport fishery for North Sea bluefin is unlikely to have caused the tunny's demise. British government scientists say that it was caused by climate change or industrial fishing of the bluefin off the African coast. This explanation remains unconvincing, as there is still a small migration of large bluefins off the west coast of Ireland and into Norwegian waters. Indeed, a bluefin of 1,241 pounds, a quarter again as large as the 1,000-pound tunny dreamed of but never landed by Mitchell-Henry in the 1930s, was caught in a trawl off the Irish coast in 2004. How much longer the run off Ireland is likely to survive is debatable now that a sport fishery has been established, exporting its catch to Japan.

More certain, and extensively recorded in the past century or so, has been the effect of fishing on what are called commercial stocks—a description that avoids considering cod, haddock, plaice, and sole as wild animals, which is what they are. The greatest impact of commercial fishing is on long-lived species that reproduce slowly, whether or not these are the fishermen's prime targets. No doubt there were once more species of shark and dogfish in the North Sea, but these are likely to have been caught as bycatch. Within the last hundred years there were common skate in British waters that when hung by the gills were the height of a man. Barndoor skate, an even larger slow-reproducing species that lives on the eastern seaboard of North America, have died out, too. In the southern North Sea, the common skate has declined by at least 99 percent since preindustrial times and is probably extinct in the North Sea, but it can take fifty years or more to prove that. It is provisionally listed as endangered by the International Union for the Conservation of Nature. Public awareness of a potential extinction, for which the public is partly to blame, is not high. Restaurants and fish-and-chip shops have long found a substitute in thornback ray, which is helpfully labeled as skate because that is what the public expects.

As to main fish species, the principal change, of course, has been in relative abundance over the past fifty years, which is likely to have all sorts of domino effects within the food web. While preparing to write this book, I had a battle on the telephone with a scientist from ICES over the question of what the virgin stock of cod in the North Sea was, or even what it was before big sailing trawlers geared up to fish for them in the eighteenth century. You would think that after a century of studying the sea at public expense, and at a point where we badly need such knowledge to make decisions for the future, the scientists of ICES would have some idea what the preindustrial spawning stock of North Sea cod was, but apparently not. Learned committees are looking into the matter and may take five years to pronounce. ICES refuses even to hazard a guess at how many cod there once were, which is convenient since a high figure would throw into greater relief the disastrous decline in cod numbers that ICES has overseen. Ransom Myers, of Dalhousie University, Canada, has estimated that North Sea cod stock was once 7.7 million tons, though some feel this is likely to ignore depredations by other species. One model said the cod, even now, would probably recover to between 440,000 and 660,000 tons if there were no fishing.

We do know that spawning stock of North Sea cod in 2005 was estimated at only 45,100 tons, so we are entitled to conclude that we have lost around 90 percent of the North Sea cod that "should" be there. Reduction to less than 10 percent of the original spawning biomass happens to be how the Canadians define population collapse. When this happened to its cod stocks, Canada banned cod fishing. The EU, on the other hand, has spent four years talking about it, and they are still catching cod. This, even the reserved, pro-fishery scientists of ICES realize, is far from sensible. But the North Sea is a "mixed fishery," with a far greater variety of commercial fish species than the Grand Banks. In other words you cannot stop catching one species without curtailing fishing for something else. In 2005, for the fourth year in a row, ICES called for a ban on fishing for cod, and severe restrictions on fisheries that catch cod in their nets as bycatch. It did this without any great expectation that North Sea politicians would be able to steel themselves to agree to such a thing. Sure enough, the EU ministers allowed a catch of about 25,525 tons, about half the entire spawning stock.

When the age of the steam trawler was at its height, the Victorian scientist Thomas Huxley famously declared that an acre of good North Sea fishing ground produced a ton of fish a week and an acre of land a ton of grain a year. Huxley meant sizeable cod, haddock, whiting, plaice, and sole landed into Hull, Grimsby, or Lowestoft. The productivity of the sea in those terms is now about a tenth of what it was in 1883 because the populations of the other fish have declined as much as the cod. Huxley fell victim to a number of misconceptions about the productivity of the sea. An acre of the best ground, of course, happened to be where fish gathered: it was the whole of the North Sea ecosystem, a much larger area, that was actually feeding them. His comparison with farming overlooks a crucial difference between farming and hunting-gathering: fishermen reap but they do not sow. Investment in new technology has the opposite effect in farming to that in fishing. An acre of Lincolnshire clay, bought in many cases by successful fishing families, would each year produce a ton of wheat to the acre in Huxley's day. It now produces over 5 tons of wheat to the acre because of advances in technology. The trajectory of modern technological hunting-gathering goes in the opposite direction. Increased fishing effort produces less catch.

The autumn-spawning North Sea herring lays its eggs on gravelly areas of the seabed, but its spawning grounds are actually located in the beds of ancient rivers, and some groups of herring still spawn in inshore waters and estuaries. This suggests that the herring once evolved to live in rivers but later adapted to the marine environment. The plentiful herring has been a source of conflict and war for centuries. The traditional drift net fishery that moved from Shetland to the English Channel with the herring shoals through the season employed huge numbers of sailing craft in the nineteenth century. The herring enjoyed a recovery during the Second World War, but its spawning stock fell from about 5 million tons in 1947 to 1.5 million tons by 1957 as the traditional drift netting was replaced by trawling on vulnerable spawning aggregations in the Channel and around the Dogger Bank, where the sheer intensity of fishing is thought to have disturbed the spawn on the gravel. By 1975, when the spawning stock had fallen to 91,850 tons, the herrings' spawning grounds around the Dogger Bank were no longer used. They are still disused today, and the autumn drift net fishery on the East Anglian coast is gone, too. The herring, however, is a rare example of the success that comes when politicians actually take scientific advice. The collapse of the stock in the mid-1970s led to a ban on fishing for four years. The stock recovered, but not to its postwar level. It started to decline again in the early 1990s, and drastic action was taken again in 1996, when quotas were cut by half. Since then the stock has been rising slowly and the trend is still upward.

With that exception to the rule, the present situation for table fish is bleak. The North Sea mackerel collapsed in the 1970s and has never come back. The major bottom-living species—cod, plaice, monkfish, and sole—in the North Sea are all now listed by ICES as "outside safe biological limits," meaning that their populations are so low that they could suffer breeding failure on a large scale. The sea is almost empty of older fish. A plaice gets no older than six years these days, but, given the chance, can reach the age of forty. A cod, likewise, will rarely reach six years of age, but has evolved to live twenty years and more, laying larger, healthier, and more numerous eggs when it does so. Fish have evolved to deal with the effect of climatic variability on their reproductive success by simply living longer, until more favorable conditions return. With fishing pressure as high as it is today, and few if any refuges in the sea, this option is closed off. Fishing pressure has even caused the cod and haddock to breed a year earlier than they did when fish survived for longer in the sea, a rare example of human-induced evolution.

Worryingly, there are other signals coming from the sea. Cod, haddock, whiting, sole, and plaice are growing less quickly than they did in former decades. The cold water copepod, a form of zooplankton that forms a part of the cod's diet, has moved 600 miles northward up the Bay of Biscay over the past decade. The question now is whether warming sea conditions—which have already led to the spread of southern species, such as red mullet and anchovy, into the North Sea—could remove fish such as cod from the food web altogether. The southern North Sea appears to have warmed by around 1.8 degrees Fahrenheit over the past twenty to thirty years. A 2005 paper by an ICES scientist, Ken Drinkwater, looked at the possible implications for the Atlantic cod of a sustained increase in temperature. He found that the threshold for increased stress on the North Sea and Georges Bank cod was 1.8 degrees F. At 3.6 degrees F both would decline. By 7.2 degrees F the cod in both seas would disappear. Interestingly, he found that the world's cod population might actually end up larger than today with climate change as cod moved northward into Arctic waters previously too cold for them. However, this expansion could be limited by overfishing. If overfishing of depleted stocks continued, there might not be enough cod to drive the predicted increase in population.

Huxley was almost right about one thing, namely, the phenomenal productivity of a shallow sea. The wanton destruction of palatable fish in their millions does not mean that there has been a decline in overall biological production. While the tonnage of palatable fish has reduced by a huge amount, the tonnage of life in the sea—what scientists call the biomass—is probably very much the same. What has changed is the relative balance between species. The sea has not been emptied, as some descriptions of overfishing misleadingly suggest. This understates the unprecedented nature of what is going on.

What has happened is that predatory fish have been removed from the system. We are not even sure whether the system will allow them back—whether what we are seeing is a one-way "ecosystem flip." The greatest weight of living creatures is no longer long-lived animals, such as the skate, cod, or oysters, but small, short-lived creatures, such as prawns, langoustines, Dublin Bay prawns or scampi, sand eels, starfish, jellyfish, worms, and plankton. Overfishing has removed many of the animals that are most attractive or delicious to humans. A few thousand tons of what is left in the sea, such as prawns and langoustines, is of quite high value. A few hundred thousand tons has its use as feedstock for salmon farms, perhaps more if plankton can be turned into food. But overall we have changed life in the sea into something less "natural," less useful, and potentially more hostile in times of crisis or disaster if we had to depend upon it, as, for example, Stone Age hunters did. Some fisheries scientists treat the change caused by such fishing as neutral and say it is for society to decide whether it is good or bad. I would say that most of society is ignorant of these things—partly the result of those scientists' failure to communicate the magnitude of what is happening to anyone except other scientists. When society has the opportunity to register its opinion about its environment it routinely concludes that it wants the environment to be as "natural" as possible.

Trawls and dredges, of course, do not just affect the species they are designed to kill. They affect the whole flora and fauna of the seabed. Without trawling, Huxley's acre of seabed would support a community of plants, plantlike animals called hydroids, bryozoans, tube worms, and a variety of shellfish, many of them important prey for fish such as cod. In the western English Channel, which is clearer, the seabed supports brilliantly colored corals and sea fans. Just north and west of the North Sea, in Scottish and Norwegian waters, there are reefs of the cold-water coral, *Lophelia pertusa*, the northern ones quite shallow, the southern ones at great depth. Some 40 percent of these cold-water reefs surveyed in Norwegian waters were found to have been extensively damaged by trawls. It is calculated that these could take a hundred thousand years to grow back. Fishing with bottom gear also leads to the destruction of shellfish and seagrass beds, maerl (ancient calcified seaweed) grounds, and fragile reefs built of tubular worm casts.

Picture a 2,000-horsepower trawler targeting plaice or sole, pulling a beam trawl mounted with rotating tickler chains across even the more featureless sands and gravels of the southern North Sea. The weighted trawl and its chains, designed to beat flatfish out of the sheltering mud, smashes everything it does not catch, particularly the burrowing animals in the sediment, disturbing them at depths of up to 8 inches. EU scientists have calculated that up to 16 pounds of marine animals are killed by beam trawls to produce 1 pound of marketable sole. Sea urchins, hermit crabs, brittle stars, and razor shells are killed or, having part of their shells broken, made vulnerable to predators. Starfish tend to be the survivors, losing a leg or two but escaping the net. The winners in heavily trawled areas are usually crabs, which move in from other areas to eat damaged creatures, and worms that take over from the easily damaged bivalves.

A study of the bivalve *Arctica islandica*, which is thought to live for up to 150 years, found that it was easily damaged by trawling. Those that were badly damaged tended to end up in the stomachs of cod and other fish, thereby reducing the population considerably. Damaged ones, however, were found to be able to repair their shells, though grains of sand became bound up in the growing matrix. By studying annual growth rings and grains of sand in the shell matrix, scientists concluded that most parts of the southern North Sea were disturbed by a beam trawl at least once a year. Some of the effects of trawling may be quite short-lived: for instance, the tracks of a heavy trawl, or scallop dredge, persist for up to two years. With some major seabed structures, such as corals, the damage is longer lasting. Many parts of the southern North Sea are thought to have once been covered with an extensive network of reefs up to 20 inches high built by the calcareous tube-building worm *Sabellaria spinulosa*. These are now extremely rare, as trawling easily destroys the reefs. A surviving example is to be found in one of the gas fields, an area where trawling is difficult because of the danger of nets becoming entangled. So what has had the most extensive impact on the seas—fisheries or pollution? I lived for more than three decades believing it was the latter, in common with the environmental groups of the day, until I walked by mistake into a lecture at a North Sea conference in The Hague in 1990. A Dutch scientist named Han Lindeboom was presenting his findings, and he claimed that the effect of commercial fishing methods on fish is clearly more fatal than pollution because we record every year the millions of tons of fish these methods kill. In general, pollution effects are local, whereas fisheries cover the whole of the North Sea—and all continental shelf seas.

Lindeboom recently updated some of his calculations. He now finds that the impact of fisheries on bottom-dwelling animals is a thousand times higher than that of sand and gravel extraction in the Dutch part of the North Sea. He also finds that the damage caused by fishing is a hundred thousand times higher than that of oil or gas exploration. The reason for these findings is that the extraction of aggregates and oil and the explorations for gas take up only small areas, whereas nowhere in the whole North Sea, other than directly under an oil rig or more recently a wind farm, is permanently closed to fishing.

•

5

MIGHTY SEAMAN

Campbeltown, Mull of Kintyre, Scotland, 6 a.m. Tommy Finn, skipper of the *Gleaner,* a 60-foot trawler based in one of the remoter parts of the Scottish mainland, steers us out into a force seven wind as it gets light. The shipping forecast says the weather will worsen, so we pass boats coming in. There are "white horses" outside the bay, so they are not going to risk rough weather. The steel-hulled *Gleaner,* with a 400-horse-power engine, can withstand more of a buffeting, though Tommy informs me that the heads (toilets) are out of commission and provides a bucket in case nature calls. Whether it is the thought of this or the smell of the fried breakfast being prepared by Hamish, the cook to the four-man crew, I already feel green about the gills. It takes me the rest of the day to find my sea legs.

I am aboard the *Gleaner* because I am curious. I have never been out in a trawler off the west of Scotland, yet I have stared out at trawlers and inshore boats for years from the Isle of Arran, where my Scottish wife's family lived and where we spent many Christmases and summer vacations. At those times, when I went for a walk up the hill, I noted that I could count six fishing boats from any high point, facing either out toward the

Mull of Kintyre or inland toward the mouth of the Clyde. How, I wondered, could there be enough fish for all these boats to catch?

As a child, I read about the sea angling festival off the Isle of Arran, the oldest and most famous event of its kind in Scotland, which attracted competitors from all over the country. Prodigious catches of skate, tope, halibut, and large cod were landed in the 1960s and early 1970s. The winning bag was a ton and a half. Over the past decade or so, the event fell into decline, the top catch amounting to only a few pounds, so it was given up. I mention this to Tommy and say it is an indication of the deterioration there has been in the Firth of Clyde, but he disagrees. As a professional who fishes all over the Irish Sea, he says there are still plenty of fish out there to be caught by those who know how. He says it is possible the amateur fishermen have just exhausted the wrecks and marks they know.

Tommy is the son of a leading fisherman and is known as a particularly resourceful and enterprising fisherman himself. In a sea where stocks of cod have virtually disappeared and where fishermen who followed the herd have gone out of business, he has turned the business of catching fish into a science. When nobody realized that haddock had returned to the Irish Sea in quantity—the principal quarry there used to be cod—Tommy turned to catching them. When he realized there was a market in fish-and-chip shops for dogfish, he caught them instead. Indeed, he was more successful than he was prepared for on one occasion. He caught so many dogfish in one haul—48 tons— that he could not haul the net out of the water, so he had to tow it into port. That picture made the front of *Fishing News*. He says he didn't lose a single fish from that haul and sold the lot for more than $70,000.

Tommy is a skipper who believes in doing things differently. Instead of using an "otter trawl," where steel doors plane through the water like kites to keep the net open, he uses a method called Danish seining. This involves dropping off a buoy attached to 1½ miles of line, then shooting the net. The boat then turns around and steams back while paying out a similar length of line attached to the net until, finally, both ends are attached to the winch, the winch is engaged, and the boat draws forward in low gear. The trawl ropes, laid out in the shape of an ellipse, dance inward over the seabed, herding fish into the trawl mouth.

Tommy explains that fish never swim outward through the disturbance of silt created by the encroaching trawl lines. I wonder how he knows, other than because it works. Since the method covers a wider area than a simple trawl, it is arguably just as efficient at catching fish as a much larger, heavier trawl. It takes less horsepower to haul, which means a big saving on fuel. Danish seining therefore minimizes a boat's costs and maximizes the fishermen's share, which is divided equally after an allowance for Tommy, the owner of the boat.

The first net is shot just southeast of the Mull, and the haul begins as the sky brightens; brilliant foam bursts past us on the fish deck and the air begins to fill with wheeling, opportunistic gannets. The gannets, by nature solitary divers, roost nearby on Ailsa Craig, the volcanic plug known to travelers as Paddy's Milestone because it marks the way between Glasgow and Belfast. The gannets have learned to home in when a vessel hauls its nets. The cod end—the toe of the net where the fish end up—comes in sight, with a respectable few dozen pounds of fish, mostly haddock, which are the intended catch. As the net comes in slowly, John and Paul disentangle the sea urchins, seaweed, and isolated fish. The cod end is opened on the deck and a miscellany of creatures spills out. The haddock are sorted into boxes and will be gutted in spare moments between hauls, thus adding value because "roundfish," as ungutted fish are known, attract lower prices. About a third of the catch is sluiced over the side. This includes fish that were the wrong size to be landed, known as "discards," and creatures with no commercial value, the bycatch.

The variety in the haul is astonishing, but nothing too exotic. There are dabs, a small turbot, a couple of monkfish, small "hounds" or dogfish, juvenile cod, and haddock under the landing size. Then there are gurnards, a dragonet, a mullet, and two small octopuses, both of which are clearly still alive and well, unlike most of the fish, which are flapping in distress with their swim bladders ruptured, and are unlikely to survive if they go back over the side. Paul puts what he thinks Hamish will want to keep for the pot into a red box rather than the white boxes in which the catch will be iced. The rest go overboard, into the mouths of hundreds of wheeling gannets. I watch, but none of the discards or bycatch survives.

Bycatch and discards are a fact of life to a fisherman. There is no fishing method that catches only the quarry. This kind of trawling is relatively clean but still troubling to someone who is not acquainted with the practice. Some fisheries and times of year produce an even larger bycatch. Some prawn fisheries regularly have a bycatch of 85 percent, though some of the palatable fish caught as bycatch will be sold. The UN Food and Agriculture Organization estimates that about a third of what is caught worldwide, some 29 million tons, goes over the side. This takes what is hauled from the sea to around 132 million tons a year. Add to that the number of organisms that are killed or damaged by net, line, or trap and are never landed—such as whales, porpoises, turtles, and birds—and the number of animals destroyed on the bottom, and the total catch by fishermen reaches something more like 220 million tons a year. Consider that much of the weight of a palatable fish is head, cartilage, bone, and offal, which goes over the side or is thrown away by processors. Consider also that about 44 million tons of fish are caught to make industrial products and food for farmed fish. Consider that some of the palatable fish caught will be turned into products for other than human consumption—as cat food, for instance. Consider that there may be an element of waste because some fish will not sell. Taking all these things into account, it is possible to conclude that the amount of protein eaten by someone or something is maybe less than 20 percent of the 104 million tons landed, and only 10 percent of the amount of marine animals destroyed annually in the oceans. These are rough figures, but, given a wide margin of error, they are about right. So catching wild fish is a wasteful business. Today, aboard the *Gleaner*, I cannot help but be impressed by the productivity of the sea.

Next we are beset by frustrations. The net comes in torn after the second haul. Although we saw nothing on the sounder, and Tommy's plotter has guided us to the exact place where he caught haddock a year ago, there is clearly some projection down there, a rock or an anchor. Tommy leaps down from the wheelhouse, grabs a needle, and directs the repairs being carried out to the net, leaving his second skipper, Lachlan, steaming to where he is to drop the next line. Then the radio picks up an emergency beacon from a fishing boat they know. It is fine weather now and the wind has not risen as expected, so there is no obvious reason why the boat might be in trouble. All the same, Tommy holds a conversation with the coast guard, then heads off at top speed on the bearing he is given. Halfway there, somebody manages to raise the boat sending the distress signal, and its skipper apologizes, saying he was below working on the engine and the beacon went off by mistake. We return to the business of catching fish, but a net snags again, and this time we have to go back to pull it off.

When Hamish's generation began to fish after the Second World War, all they had to find the fishing grounds were compass coordinates on the back of a cigarette pack. Now, using global positioning system (GPS) satellite information, Tommy plots three more trawls on lines that were successful at the same time last year. Since the end of the cold war, the GPS satellite, designed originally for military and naval use, has become even more accurate because the United States has improved the resolution of the signals. This means there are fewer no-go areas for boats, and they can fish within less than 33 feet of known rocks and obstructions. For most of the day I am out with Tommy, there is also a comforting gray cloud on the bottom on the sounder screen, an indication of fish.

Now the crew is rapidly gutting away on the fish deck, and a moment of decision is upon us. When you add today's catch to what has been on ice in the hold overnight, there are nearly eighty boxes of cod and haddock on board. Tommy holds a consultation on his mobile phone, hears that prices are holding up in Fleetwood, and decides to head for home rather than stay out another four days, as originally intended. A white van is waiting as we tie up at the quay. We are all looking forward to an unexpected meal ashore, a drink, and a night in a bed rather than a bunk.

.   .   .

That was nine years ago. Tommy is still fishing, mostly with the same crew. There is not much else to do in Campbeltown, except for working in the seasonal tourist trade or for a new company making wind turbines. The name *Gleaner* now belongs to a bigger boat—80 feet—that Tommy acquired a couple of years ago. He says there were fish in the sea and he thought it worth gearing up. He sold the old boat for a good price, and the bigger one was cheap because so many people were trying to get out of fishing. He is pair trawling now for haddock with another vessel. They are doing okay, but the quotas are currently a third of what they were nine years ago. There is a ban on fishing for cod in the Irish Sea, and scientists say drastic reductions are needed to save the haddock. Increasing amounts of what fishermen land is illegal. Someone Tommy knows just landed a thousand boxes of haddock in Dublin, all illegal. Skippers keep no records and sell without documentation—probably for less than the going rate—to keep their cash flow positive and pay the mortgage on the boat. Tommy is philosophical about everything except the prices he is getting for his fish. "The scientists say there are no fish, but you never see a scientist in the Clyde when the cod are there in March and April. Come those months you will find thousands of boxes of cod out there." There are times when some fisheries do well; this is a time when his is not. The pelagic boys who go out for mackerel and herring are doing well on rising quotas at the moment. What gets Tommy is when he does his job well and he and his partners catch hundreds of boxes of precious, legal haddock: even then, ungutted fish fetch only $35 a box when they should fetch $70, and gutted haddock fetch $53 when they should fetch $106. Tommy blames the price on imports from the Faroes and Iceland, where there is plenty of fish.

I haven't the heart to tell him my theory, which is that the domestic market is down because so many of his fellow fishermen are cheating—catching far more than their quota of what scientists now regard as endangered stocks of fish.

There is a woodcut on the wall of a fish-processing firm in Plymouth, England, that I study as I wait to meet the owner. It features salty, idealized fishermen in sou'westers hauling a hemp net loaded with fat fish into an open sailing boat. It could be the Sea of Galilee in biblical times. It occurs to me that there is something timelessly right and something annoyingly wrong about this stylized image of fishermen. I struggle to define what irritates me about this picture. I come up with this. The picture tells the old truth that the real price of fish, the price you don't pay over the counter, has always been reckoned in men's lives. (Even now, in this age of gender equality, the fish-catching industry remains dominated by men.) On average over ten years a British fishing vessel has been lost at sea every twelve and a half days. There is a page of wrecks in every month's *Fishing News International*.

But now the real price of fish has to be reckoned in a more complex equation: danger to fishermen versus damage to populations and ecosystems on which humans ultimately depend. The world has changed. This is what reaching the end of the line for so many fish stocks means. The depiction of strong men risking their lives as they battle against the elements isn't wrong. Yet the odds on the survival of the hunters have increased vastly since the days of steam trawlers, semaphore signals, and nineteenth-century trawl captains who made cabin boys transship the catch at sea in open rowboats. Safety increases by the decade, and who would not wish it to? Fishing is

still an unenviably dangerous business, with poor working conditions made worse by bad weather, and requiring courage in
autumn and spring gales, in tropical storms, and aboard long-
distance fleets that fish near the poles. The dangers of fishing
are real, but fishermen relentlessly trade on this to their advantage. An old friend, a civil servant and press officer, once told
me after bruising annual negotiations with Scottish fishermen
and TV coverage that gets politicians hopping mad: "Just when
we've got those buggers where we want them, someone starts
playing 'For Those in Peril on the Sea.' "

The risks to fishermen remain real, but our automatic sym-
pathy for them has been modified in our times by a new truth
that has yet to find equilibrium: the odds on the survival of the
hunted have been massively reduced. The development of fishing technology means that no area of the sea is inaccessible to
nets or long lines. It used to be said of good Icelandic trawler
skippers that they thought like fish. Now they can see like them
as well.

The World Fishing Exhibition, Vigo, Spain. Advertising slo-
gans for the latest developments in engines, boats, nets, lines,
satellite phones, GPS plotters, sonar screens, and computer systems for analyzing what all of these are doing compete for attention. Visitors to these several acres of covered exhibition
floor right by the airport are urged to "get the ultimate weapon
on board" or to buy "the world's largest trawl net" or to keep
control of an entire fleet of satellite-connected, fish-tracking
buoys through a "mobile Earth station." A hit this year is the
New Zealand stand advertising the sea-mapping software Piscatus 3D (slogan: "Fish hate us"). Selling technology to fishermen is a macho business, rather like selling pesticides to farmers

in the boom years of the 1970s, when top brands were given names such as Rapier, Impact, and Commando. Where pesticide manufacturers tended toward images of combat to promote their products, fishing industry copywriters seem to prefer the language of science fiction or wars of mass destruction. Legal, aboveboard public companies sell technology to fishermen, not all of whom will use it legally. Every major fishing gizmo supplier and boatbuilder and some of the big players in fishing have decided to be seen during this festival, which takes place every three years, drawing sellers and buyers from more than thirty countries in Europe, Africa, and Latin America. Some, such as the giant Spanish company Pescanova, aren't really here to sell anything other than themselves. They are here to polish up their global image. Pescanova has a huge stand with hardly anybody in it, featuring video displays instead. Galicia's politicians line up to speak at this exhibition, and an amiable dinosaur from the Franco era, Manuel Fraga, who is in his eighties, speaks at the opening. The Spanish fisheries department needed no encouragement to sponsor a stand; amusingly, it confirms many people's impression of their countrymen by sporting posters in Spanish encouraging fishermen not to land undersized fish. Even the European Commission has been induced to take a stand, a sign of Spain's mysterious clout with the fisheries directorate general in Brussels.

Star of the show, in terms of the attention it receives, is the plotter, a flat-panel, high-resolution computer screen, perhaps 18 inches square, now the centerpiece of any large freezer trawler's bridge. Most trawlers will have two—this year's model and last year's model—in case one goes on the blink. Fishermen don't use charts anymore—not at this level, anyway. The plotter brings together navigational data from all the onboard equipment onto one screen in front of the skipper's chair, with the contents of every other screen on the bridge just a mouse click away. Several versions are on sale by MaxSea (France), Simrad (Norway), and Furuno (Japan), together with screens for every other conceivable purpose. There are developments of the plotter, depending on what you are fishing for. For tuna purse seiners and other pelagic boats, the MaxSea salesman tells me, the latest thing is to integrate navigational information with sea temperature contours from satellite weather channels. This enables the skipper of the purse seiner, a sizeable ship that might carry its own helicopter or speedboat, to find the thermocline, the abrupt edge between masses of warmer water and cooler water where the dynamism of the sea is greatest. Here plankton are generated and prey fish congregate, to be picked off by milling whirlwinds of tuna.

These feeding frenzies, in rainbow colors on the screen, are picked out in simulations run on the latest forward-facing, low-frequency sonars in the exhibition hall. One shows a feeding frenzy of tuna 2½ miles ahead. Downward-facing, split-beam sonars, represented by other screens, show what is happening in the sea below. Tuna fishermen also use banks of bird radar, on sale here as well, which are tuned to a pitch more sensitive than those used to detect aircraft, in order to pick up the flocks of birds that mark tuna shoals. Powerful, space-age pier binoculars, four or five times the size of those used by U-boat captains, are mounted on the bridge or in the crow's nests that many of these vessels have. These are used to spot the free-swimming shoals that the purse seiner's speedboat will race to enclose within the net.

Tuna fishermen know that tuna congregate around floating objects, having observed them around rafts or logs. At first they set their nets around these rafts to catch fish. Now each boat makes its own fish aggregation devices (FADs)—dozens of them, or as many as they can monitor. The FADs can be equipped with an echo sounder to monitor fish activity. They also monitor water temperature and speed of drift to assess whether the buoy has reached a thermocline. Each boat launches its FADs into the ocean currents, which in the Indian Ocean begin around Madagascar, and recovers them about 1,500 miles to the northeast, around the Chagos Archipelago. FADs used to have radio beacons, and many still do because they are cheap, but the falling price of satellite telephone technology means that the latest gadget to have is a buoy linked to the Inmarsat satellite. This sends a signal to a control center that is capable of monitoring data from up to 600 FADs. If fishermen can afford a dozen or so buoys at $225,000 each, it goes without saying that successful tuna purse seiners are extremely profitable.

When it comes to trawler technology, the latest development in an industry that progresses as fast as the software industry itself is seabed-mapping software, such as Piscatus 3D. This combines modern computer technology with the traditional echo sounder to extract even more information from the sounds it sends to and receives from the seabed. The result is that the fisherman can see into the depths in virtual reality. So that's why fish hate them. The information packet from Piscatus explains: "We have developed a comprehensive 3-dimensional fishing tool that shows you exactly what is happening as you fish. See your boat, the seabed, the fish, your gear: and even door-spread [the distance between the planing doors holding the net open] in a real-time, animated 3D landscape." It's a cross between *Star Wars* and Flight Simulator. This is the kind of stuff that fishermen buy when they are operating in waters that haven't yet been overexploited and where fat profits are still to be made if they get there first. Equipment like this makes most trawl fleets in domestic waters look like cottage industries.

All this innovation can be puzzling at times to the tattooed, weatherbeaten men who come to exhibitions such as these. But the software people are not going to let puzzlement stand in the way of profit. "The only ones frightened by our technology are the fish," says the brochure put together by Piscatus's parent company, Seabed Mapping. "As a skipper . . . you want to spend your time catching fish and not using computers. Seabed Mapping is part of the fishing industry and has developed Piscatus 3D alongside world-leading skippers. As a result, it's both highly effective and extremely easy to use. . . . Being out of favor with the fish and making more money has never been this easy." MaxSea's personal bathymetric generator (PBG) does something very similar, allowing fishermen to create much more accurate 3-D maps of a seabed. Piscatus says its mapping software allows you to "let the net drive the boat." You can even fly your net down the line of a previous tow. Fisheries experts have spent decades debating how to quantify what they call "technological creep," the ability of a single boat to catch more each year because of improvements in technology. What we are looking at here looks more like technological gallop.

For trawlermen fishing in deep water in the North Atlantic, the Indian Ocean, the Tasman Sea, or the Pacific, underwater mountain tops, known as "sea mounts," can be a particular hazard to nets. Consulting a 3-D picture of an area of seabed before and after MaxSea's PBG has been over it is the difference between looking at gently undulating prairie and the Rockies—the latter being an accurate description of the pinnacles and chasms on a rocky seabed. The publicity material explains:

The great advantage . . . is that it enables you to explore areas that would otherwise be too challenging—where other fishermen will not venture. Areas that have not been overfished and so promise particularly abundant catches . . . The unmatched accuracy . . . means you can go hunting for those "hidden hideaways" (shelves, canyons and crevices) hitherto invisible and so virtually never fished.

A glowing endorsement follows from Michel Derosière, skipper of the *Fils de la Mer*. He says:

It's as if the water had been drained away and I can look right down and see exactly what the sea bed looks like. In the beginning I only got a fairly approximate view of the Channel floor. But now . . . it's fantastic. It means I can update the data in real time—so we can fish areas we used to avoid, whatever there might be down there, even the most treacherous shelves or rock formations. And we always know exactly where we are—to within a meter. This is a great new tool which soon pays for itself because there aren't many others fishing where we go, so we can be sure of hauling in much bigger catches.

One shudders to think that he's probably talking about the English Channel. This is fishing technology outstripping the speed at which most people can comprehend what fishermen are actually doing, let alone figure out whether they approve of it. Rather like a car salesman who lures the gambler into the showroom with the latest Bentley or Mercedes parked out front, the MaxSea salesman points out that a lot of his company's fourteen thousand customers are skippers of small boats, and he has a pitch for them, too. He senses I'm probably not going to buy the top-of-the-line PBD, the "ultimate fishing aid," and starts a different tack: "You can map a big rock. You can go to 1.5 meters [5 feet] resolution. If you have lost your gears, you can get them back."

The extent to which computer-based information technology has improved the fisherman's chances was described to me at a conference about developments in the deep sea held in conjunction with the exhibition. Halli Stefanson, an Icelandic fisherman who emigrated to New Zealand and now skippers a 2,750-ton, 280-foot ship catching orange roughy, explains that these deepwater fish tend to congregate in a cloud on the tops of sea mounts. The process of catching them, with the latest computerized trawl-monitoring equipment, works like this: "You can just drive the net on to the hill. It is a revolution for us. You try to drop the net as close to the top of the hill as possible and drop down the side. If you are lucky, you get about five minutes' fishing. This is typical of the orange roughy fishery. But you can get 17 tons as a result of two minutes' fishing."

Acoustic net monitors, known as "suitcases," show where the net is. Catch indicators, or "eggs," show when the net has 6.6, 11, 44, or 66 tons of fish in it. Net monitors are useful on any net, but they are virtually essential when managing one of the giant Gloria midwater trawls, the biggest nets in the world, used to catch redfish on the mid-Atlantic ridges. The redfish is a solitary swimmer and does not shoal, hence the need for a huge net. The latest, on sale in Vigo, has a mouth opening of 43,000 square yards, large enough to catch half a dozen 747s flying in formation. These nets are six times larger in the opening than the original Gloria trawls of ten years ago, but this does not mean fishermen want to fill them up before they haul. Too many fish means mangled fish. A skipper in Reykjavik told me that he uses the net indicators to haul when there is a maximum of 6.6 tons in the net so that the fish are top quality and very fresh. If you go over that amount, the fish tend to get damaged and aren't worth as much. Alaskan pollock fishermen tell me they use net indicators for the same reason. Their fishery, which is better managed than many, prizes quality before quantity.

The potential of technology is, of course, neutral. It can be wasteful or clinical, depending on how it is used. It can be the tool of the police officer or the poacher. "Blue boxes," which are satellite transceivers that transmit a position several times a day to tell regulators where fishing vessels are situated, are now required under EU law and in many other jurisdictions. They are, for example, required for tuna boats fishing under license in Kenyan or Madagascan waters. The only trouble, I was told by Luis Díaz del Río, the director general of Satlink, a firm that makes all sorts of satellite links, is that Spain, Portugal, and Britain all require their vessels to have different sorts of blue boxes. The Spanish one, surprisingly, sends more information than the British version. Some boats have all three if they are planning on spending time in all three countries' jurisdiction.

There is even a technical possibility of managing trawling to make it less indiscriminate. Fisherman could sense the size of the fish they are going to catch before they catch them, or set up a trawl to skim the bottom, causing minimal damage to the seabed. It goes without saying, however, that it tends only to be the developments that increase the catch that are used by fishermen because there is no incentive to do otherwise. Here is Halli Stefanson again on the revolution he has seen in guiding the net over the past five years.

This is all well and good, but recent information shows that sea mounts are very vulnerable to trawls. We must be careful how much we take out. Unfortunately, the mindset of many people in the industry has not changed since pre-technology days. Most sea mounts only yield fish for a year or two in any reasonable quantity. Orange roughy rely on a certain amount of biomass being present before spawning takes place. Bottom trawling can be disrupting the spawn for 3–4 years. In Iceland, where conservation practices have been in place since the 1860s, we fish on a sustainable basis. If we are to fish on sea mounts, we need to change our attitudes. Unfortunately, I'm not that optimistic. We had an agreement to protect sea mounts on the high seas in 1999 but it didn't work. Unscrupulous individuals exploited the loopholes. They made a lot of money, but the fishery has never been the same since.

In another session at the conference Grimur Valdimarsson, its Icelandic chairman, delivers a blunt conclusion to the deep-sea debate on fishing techniques. "Technology," he says, "now makes it possible for us to catch whatever fish we like." Malcolm Clarke, a fisheries scientist with one of the privatized fisheries bodies that now assess orange roughy stocks for New Zealand's companies, puts it another way: "Our understanding of how to exploit the resource has moved much faster than our ability to manage it."

6

THE LAST FRONTIER

*The oceans are the planet's last great living wilderness,
man's only remaining frontier on Earth, and perhaps his last
chance to prove himself a rational species.*

—John L. Culliney

The fish consumers are most familiar with come from the shallow seas of the continental shelves or the surface waters of the open ocean. But there is plenty more water out there with fish in it before you get to the abyssal depths. As fish stocks began to decline the world over, enterprising fishermen began to look at opening up a new frontier in deeper water. In the Atlantic, that frontier was already being explored by the oil industry. The technology to fish at depths beyond 1,000 feet is relatively new, part of the rapid development of boats, engines, winches and electronics that has happened in the past thirty years or so. The fish of deep water—the constantly dark world is generally agreed to begin 1,300 feet down—are different in many ways from shallow-water fish, and the regulations governing how fishermen exploit them are in their infancy within 200-mile limits and nonexistent elsewhere. The concerns about their exploitation are global and just as troubling as concerns about the use of shallower seas.

Before we descend to the dark world of true deep-sea fish, we must not overlook the unfortunate species, such as ling, tusk, Greenland halibut, and blue whiting—once regarded by fishermen as deepwater fish—that no one seems to want in their scientific categories or management regimes. These are the fish of the continental slopes and midoceanic banks, which tend to find themselves just on the edge of countries' territorial waters. As such, they are a cause of serious diplomatic trouble. Greenland halibut, or turbot, was the source of a major dispute between the EU and Canada over the international waters off the Grand Banks in 1995, and complaints about overfishing by Russian and EU vessels flicker on to this day. But now the hot issue is blue whiting.

Blue whiting, a member of the cod family, is found in waters from the surface down to 3,300 feet but is most often caught at 660–1,300 feet. It reaches sexual maturity between two and seven years of age and can live for twenty years. Blue whiting ranges from the Barents Sea north of Finland to the mid-Atlantic around Iceland, and southward as far as North Africa. In the west Atlantic it is caught in southern Canada and along the northeastern coast of the United States. It is a cousin of the whiting but lives at greater depth, behaving like a pelagic fish of the deep sea, living near the bottom by day and making daily vertical migrations to the surface at night.

Surprisingly, for a fish that few people have heard of, blue whiting is about fifty times more numerous than the remnant North Sea cod, and nearly as numerous as Alaskan pollock, the largest remaining stock of palatable fish in the world. So blue whiting is precisely the sort of fish that fishermen turn to when more accessible stocks run out. The vast proportion of it is never seen on a plate but instead is processed into oil and fish meal. I went to enormous lengths to get hold of some to see what it was like and persuaded a well-known British seafood chef, Rick Stein, to cook it. The fillets were small but with a nice texture. They tasted surprisingly good. There is at least one Icelandic chef who says that blue whiting is an excellent table fish. Something or somebody is eating blue whiting in large quantities, because catches of it have trebled in the past decade. There is, I am told, some demand for blue whiting as a palatable fish in Russia and the Baltic states, though I have never seen it sold, but mostly blue whiting is in demand because it makes excellent food for farmed carnivorous fish such as salmon. As other sources of fish meal for aquaculture falter, the blue whiting has been enjoying an exceptional period of breeding success.

So a gold rush is in progress. Catches of blue whiting in the eastern Atlantic have risen from an impressive 731,320 tons in 1987 to a staggering 2.6 million tons in 2004. Scientists from the International Council for the Exploration of the Sea (ICES) in Copenhagen have said consistently that a sustainable catch is somewhere between 715,000 and 1.1 million tons. The worst culprit is Norway, which is catching nearly 880,000 tons a year, virtually the entire sustainable quota, for itself. Vast trawlers are using giant Gloria trawls at depths of around 1,650 feet to catch huge quantities of blue whiting and are steaming back to port with the waves breaking across their decks because they are so full. No one seems to be in a hurry to stop this carnage, which can only end in the collapse of the blue whiting stock. Catches have already fallen by 20 percent. After three years of diplomatic impasse, politicians of the European Union, Norway,

Iceland, and the Faroes finally agreed in autumn 2005 to set a quota for blue whiting: 2.2 million tons, to be reduced by 110,000 tons a year in succeeding years.

I asked Kjartan Hoydal, secretary of the Northeast Atlantic Fisheries Commission, which attempts to control fishing in international waters in the eastern Atlantic, what seemed to be the problem. He replied: "With blue whiting it is obvious what should be done, but nobody seems to want to do it. The fishery could be one or two year-classes [years when spawning survival was good] and that's it."

The prognosis for the fish stocks of the deep sea—that is the dark depths from 1,300 to 10,000 feet where fishing is now feasible—is, if anything, bleaker. Although we know that deep-sea fish have been caught in quantity for more than twenty years, we are otherwise rather ignorant about them. We do know that in the entirely dark world below 3,300 feet, fish tend to be long-lived, less numerous, and slow-growing. They are either scavengers, waiting for infrequent "food parcels" from the brighter regions above, or, like the many species of deep-water shark, predators that eat the scavengers.

The overall abundance of deepwater fish is much less than the fish of shelf waters, and catching them would be a costly, unprofitable business if it were not for their tendency to congregate around geographical features, such as raised banks and ridges, for feeding and spawning. Most dynamic and productive of these deep-sea habitats, but little known, are sea mounts, former volcanoes that rise from the ocean floor. Where sea mounts break the surface, such as in the Azores and Hawaii, we know them already as midoceanic islands. There are estimated to be thirty thousand sea mounts in the Pacific, but only six thousand in the Atlantic, and only one thousand have been named worldwide. Far fewer have been studied. As George Clement, managing director of Seabed Mapping, the New Zealand company that manufactures the Piscatus software that compiles 3-D maps of the seabed, put it: "The sea floor is less well mapped than the surface of the moon."

The strong localized currents and upwellings created by sea mounts make them hot spots for plankton, and consequently for small fish and the large fish that prey on them. Scientists tell us that sea mounts, oases of diversity in vast expanses of open ocean, are likely to support large numbers of undiscovered species. They certainly support communities of suspension feeders, corals, sponges, and sea fans that filter organic matter from the water rushing over the summit. Orange roughy, among other fish that gather to spawn on the top of sea mounts, feed on prawns, squid, and small fish that float by. Further down a sea mount the coral becomes less dense, and lobsters and sea spiders hide among the rocks. Unfortunately, the opportunity to study even a representative selection of the world's sea mounts before trawling fundamentally alters them is fast running out.

While sea mounts are the most spectacular habitats, there is enough diversity in the species that live in the microclimates caused by ridges and currents on the continental slopes of the Atlantic basin to keep a biologist busy for a lifetime. Dr. John Gordon is an honorary fellow at the Dunstaffnage Marine Laboratory, on the west coast of Scotland, where he worked as a biologist until his retirement. Now he works there anyway, mostly unpaid. He began to study deepwater, bottom-living fish when this was regarded as an extremely esoteric subject. He saw his first catch of deepwater fish on the research ship *Chal-*

lenger's voyage to the Rockall Trough in 1973. This area is now the center of intense fishing activity by French and Spanish trawlers, which were the first to specialize in the deep sea as shelf seas became overfished.

Gordon now finds himself a lone voice in warning that the fisheries of the Rockall Trough and Hatton Bank are being subjected to unsustainable fishing pressure—in other words, they are fast being mined out. In 1997 a new British government decided to ratify the UN Convention on the Law of the Sea. It therefore renounced its claim to a 200-mile limit around Rockall, a granite outcrop about 80 feet high 200 miles to the west of the Outer Hebrides. The reason given was that international law now views claims based on uninhabited rocks as untenable. However, this does not prevent other countries continuing to inhabit them occasionally, and defending *their* claims. The consequence of this decision for fisheries does not seem to have been considered. It means that much of the Rockall Trough, an area of deep water that has been studied since the 1860s, and the whole of the Hatton Bank are now in international waters. As a result, they have been the subject of a fishing free-for-all.

At the conference in Vigo, where Halli Stefanson told us about the excitement of fishing the deep sea with the latest technology, Gordon explained the differences between the fish of shallow seas and the weird and little-understood animals of the deep. Between 1,300 and 6,600 feet down, he told us, there are more than a hundred species of bottom-living fish. Deepwater fish have either deciduous scales or none at all, making them more vulnerable to damage by trawls, even if they escape through the mesh. Such fish, including orange roughy, scabbard fish, smoothheads, rabbit fish, scorpion fish, and grenadiers, tend to have large heads and tapering bodies, so undersized fish get caught in nets that would allow the young of differently shaped shallow-water fish to escape. There is no wave action at depth, and such currents as there are do not generally have the same force as those in shallower seas, so lost or snagged fishing nets tend to go on fishing, often for decades, as "ghost nets."

The other reason why fishing has a greater impact upon deep-sea fish is age: some of these fish are remarkably long-lived. The orange roughy, for example, is now assumed by most scientists, including Gordon, to live up to 150 years. It takes twenty years to mature and does not reproduce until it is about thirty. (It must be noted that there is still some controversy about dating old specimens of orange roughy. Based on growth rings found in fishes' ear bones, like those found in trees, dating assumes that one ring equals one year, which would make some of the roughy that have been caught very old indeed. Some scientists and fishermen, however, are convinced that the rings do not represent years and that the top age of a roughy will turn out to be about eighteen.) Even when it does mature, the orange roughy is less productive than shallow-water species. It lays only tens of thousands of eggs, an order of magnitude less than the cod's several million.

A similar pattern of long life and low fecundity is found in other deepwater fish. Round-nosed grenadiers mature at eight to ten years old and live to seventy-five. Smoothheads, such as Baird's smoothhead, live to thirty-eight years, and the leafscale gulpher shark is known to live to seventy and to have only six to eleven live young. Large numbers of these fish are known to be caught and discarded by long-line vessels targeting other species, but few records are kept. The effect of fishing on these nontarget species, just like their basic biology, is almost completely unknown. Gordon was surprised when he went aboard research ships carrying remotely operated survey vessels (ROVs) to find that the images they beamed up most frequently from the depths were of fish that did not get trapped by trawls at all. The most numerous was the cutthroat eel *(Synaphobranchus kaupi)*, which is thin enough to survive being caught in all but the most fine-meshed nets. The cutthroat eel is the most abundant species in the Rockall Trough and the Porcupine Sea Bight, off southwest Ireland. Gordon found it was also in the Bay of Biscay when he was invited on to the French vessel *L'Atalante* in 1992.

It is inherently surprising that anyone wants to eat a fish like orange roughy that, as far as we can work out, outlives even the longest-living humans. It is known as the Queen Mother of fish. Just as remarkable is the fact that the main European market for this fish, which produces firm, bland, but relatively tasteless white fillets, is France, a nation renowned for its demanding culinary standards. France began catching blue ling in the 1970s, and started on orange roughy in the North Atlantic in 1991. It is now the leading country in Europe for catching and eating deepwater fish. How did this come about?

The unattractiveness of many deep-sea fish was overcome by filleting them, which makes them look like any other white fish fillets. Then, French marketing people tackled the problem of unfamiliar names, and here they succeeded beyond their wildest expectations. Abandoning the dull common or scientific names the fish were encumbered with, the marketing people equipped them with dashing, military-sounding names from a glorious Napoleonic past, designed to sound a note of pride in French breasts. Black scabbard fish became *sabre*, orange roughy *empereur* (which happens to resemble the name for swordfish in Spanish, *emperador*), and round-nosed grenadier became simply *grenadier*. At the time the new species were reaching the market in the early 1990s, the processing industry was crying out for more white fish fillets, so once the marketers had solved the naming problem, the niche was filled. Orange roughy also finds ready markets along the West Coast of the United States and in Chicago and the Midwest, where its meaty fillets are sold as the fish for people who don't really like fish.

The effect of trawling in deep water in the North Atlantic has been to reduce all known fish populations there to around 20 percent of what they were in the 1970s. That is true of predator species, such as round-nosed grenadier, black scabbard fish, and sikis (a kind of deepwater shark). With orange roughy, the attrition appears to have been worse. The fishing in British waters started in 1991. By 1994 the catch rate was 25 percent of initial catch rates. Effectively, the known orange roughy populations of Hatton Bank, Porcupine Sea Bight, and Rockall Trough have been mined out within a few decades of being found. Stocks of round-nosed grenadier west of the British Isles are now so low that they are below the precautionary level set by the ICES—the level at which fishing should theoretically stop. For most deepwater species a sustainable exploitation level is around 2 percent of the original stock, compared with 20–30 percent for shallow-water fish. Scientists say it may simply not be possible to harvest a deep-sea stock sustainably—that is, take a commercially viable number of fish every year. All that we can perhaps do is harvest them down to a certain known level from which we hope they will recover, then leave them alone. This assumes we know enough about the biology of the fish to know the level at which the population can recover, and can measure it. It also assumes we have control over the resource, which is doubtful considering that international waters are managed by the Northeast Atlantic Fisheries Commission, which controls the Hatton Bank. The gloomiest conclusion came in the mid-1990s from Phil Aikman, author of a Greenpeace report on deep-sea stocks. He said we should not be exploiting deepwater fish at all. The one country that claims it has finally found a way of fishing orange roughy sustainably is New Zealand. Its exclusive economic zone (EEZ) is mostly deep water, which includes extensive ridges and sea mounts, so it got into deepwater fishing twenty years ago. There was a time in the 1980s when New Zealand fishermen were catching 100,000 tons a year of orange roughy, a significant proportion of the world catch. Then scientists found that stocks were a fraction of the size they thought they were, so fishing had to be scaled back. The difficulty was in estimating the stock size.

Scientists—who in New Zealand work for private companies and are hired by fishermen, who pay the cost of research—differ in their opinion of the stocks even now by a factor of three. Nine out of ten roughy stocks are not looking good. Some are closed to fishing altogether, and only one appears to be rebuilding, but even that could be because the form of scientific assessment has changed. Malcolm Clarke, who works for a privatized fishing science company, told the Vigo conference: "Orange roughy has an uphill battle when it comes to sustainability. Our overall experience in New Zealand doesn't look good. A lot of management lessons have been learnt."

George Clement, who is managing director not only of Seabed Mapping but also of the Orange Roughy Company, which manages the fishery, says the orange roughy is now being fished sustainably after many years of scientists being overly optimistic about how many fish were there. Clement, a believer in the theory that the orange roughy reproduces long before it is thirty and lives to nothing like a hundred years old, believes some stocks are actually rebuilding. Unfortunately, it will be ten years or more before we find out if he is right—by which time a lot of more mature roughy will have been caught.

In Europe and North America, there doesn't seem to have been the same overassessment of deep sea stocks that happened in New Zealand. Instead, in the eastern Atlantic there has been bureaucratic inertia in the face of a whole new area of the sea to research and control. Repeated warnings have been sounded, but the European Commission has reacted far too late. In 2000 it announced quotas for deep-sea stocks, but these are far larger than recommended by ICES. The best summary of the situation in the waters off the British Isles now being fished mostly by France and Spain is provided by John Gordon:

> There is general agreement among scientists, the fishing industry and the politicians that the fragile, deep-water stocks are seriously over-exploited, but political imperatives dictate that uncertainties and inconsistencies in the scientific assessment and advice are used to postpone the urgent action that is required. It is perhaps not much of a consolation, but at least in the Rockall Trough, we know a lot about the ecosystem that is being destroyed, while in other areas, such as the Hatton Bank, we will never know what is being destroyed.

He's right. To our children and grandchildren it will sound like slim consolation.

THE INEXHAUSTIBLE SEA?

Lowestoft, England. January. Six hundred years of enterprise directed at pulling fish from the North Sea in all kinds of weather ends like this, with the unrepaired doorways and shabby 1930s office buildings on the seafront telling a story of economic collapse. So, too, does the fish dock of what was once one of England's greatest fishing ports, famous the world over. The gray-brown swell and the driving rain cast a mood of gloom, which is worsened by the gaping crack in the wall and the flapping, broken panel of the company secretary's desk in the offices of Colne Shipping, the last company to run a fleet of beam trawlers fishing for sole and plaice on the town's traditional fishing grounds off Dogger Bank. Once every British schoolchild knew that the Dogger Bank was the North Sea's most productive fishing ground and that Lowestoft was East Anglia's most productive fishing port. Hugh Sims, company secretary of Colne Shipping, explained how in August 2002 the company made one of its hardest decisions in its fifty-eight-year history and tied up its fleet of 130-foot trawlers for the last time because the price of fuel had gone up and the number of fish had gone down.

Inshore boats still go out for a night and land a box or two of cod, rays, and dogfish, plying a trade not dissimilar from when Lowestoft was founded as a fishing village in the fourteenth century. There is a run of sprats, someone has discovered, which the Danish sand eel trawlers with their fine mesh nets haven't vacuumed up yet. But the deepwater fleet has gone. Lowestoft lacks a future, except as a retailing and commercial center for the regional tourist trade, for there's not much to see in Lowestoft. The terrible sadness that seeps out of the place is reminiscent of a mining town where the coal ran out.

When in 1965 Hugh Sims answered an advertisement for an accountant placed by the Boston Deep Sea Fishing Company, Lowestoft was a thriving port with 120 fishing vessels in the dock. There were five fishing companies, each of which employed its own staff—skipper, crew, and onshore workers including fitters, mechanics, and icemen. Only your own staff would get up at a minute's notice at 3 a.m. so you could catch the tide. The whole town, from the prostitutes in the dockside bars to the proprietors of prosperous shipping companies, celebrated enterprise and extravagance. Successful skippers wore the port's traditional shore leave suits in garish colors ranging from red to lime green when they went out on the town. When one successful skipper bought a Mercedes, the managing director of Colne Fishing, as it was called then, said, "Good. Now the others will want one."

From its heyday in the 1950s well into the 1980s, Lowestoft was the principal supplier of plaice and sole to Billingsgate, London's fish market. Dover sole was a misnomer. The fish boned at your table in a fancy restaurant was more likely to come from the Dogger Bank via a train from Lowestoft. Competition led the five Lowestoft companies to consolidate and cut staff. They moved to fewer, larger trawlers that could stay two weeks at sea. New techniques of netting were tried. Then in 1984 Colne Fishing moved to the Dutch style of beam trawling, using powerful 2,000-horsepower trawlers dragging nets weighted with heavy beams and tickler chains along the bottom to force the sole, plaice, and monkfish up into the net. Beam trawling was a highly efficient way of killing fish. The only disadvantage seen at the time was that it was heavy on fuel.

Profits declined slowly through the 1980s and 1990s. The best skippers saw there was little money left in the southern North Sea, so they moved away, to the south coast of England or the west of Ireland, where their paychecks were larger because they caught more fish. In the end, Colne had to accept that its vessels were too large and too expensive to run for the amount of fish that were left to be caught.

The irony slaps you in the face like the January gale. The fish dock is all but empty, the herring drifters gone, the last beam trawlers that sailed from here broken up or sold to flail the bed of some other ocean, but the government laboratory that was set up to ensure the survival of a plentiful supply of fish lives on, dominating the town, costing taxpayers money, monitoring a sea that it was supposed to save. The laboratory's new 240-foot research vessel, the *Endeavour*, which cost $43 million, dwarfs other boats in the dock. Yet despite all this public expenditure we have a sea where the stocks of two-thirds of the major commercial fish species are "outside safe biological limits." That is the accepted scientific way of saying that they are close to the point of no return.

The central problem for fisheries science is counting fish. You cannot, even with current technology, see how many fish there are in the sea, so you have to find other ways of counting them. So did scientists around the North Sea get their numbers wrong? Or were their endeavors routinely overwhelmed by those of politicians and fishermen with different agendas? Or are scientists effectively just another interest group, with only a marginal interest in achieving the conservation of fish on their path to career advancement? Such explanations have all been offered at different times and on different continents for the manifest failure of modern science-based management, with all but a few exceptions, to conserve fish.

Not being a scientist, I find myself frequently annoyed by the vast amount of bureaucratic language churned out by the majority of fisheries scientists. They might believe they are being neutral and nonjudgmental, but they are in fact betraying a clear bias in favor of large-scale industrial fisheries as the principal use of the sea. To avoid being shaped by their patterns in the sand, it is necessary to tell the story in a different way, as history, through the lives and perspectives of a few individuals.

The Lowestoft fisheries laboratory is housed in a redbrick building, formerly the Grand Hotel. Today it is part of a government agency with the forgettable name of the Centre for Environment, Fisheries and Aquaculture Science (CEFAS), but it is popularly known as the Lowestoft Laboratory. Its deputy director and senior scientist, Dr. Joe Horwood, is a cautious man who seems temperamentally averse to making strong statements. Yet it has been his job to issue regular and unambiguous warnings to fishermen and politicians on the state of the dwindling shoals of cod, haddock, plaice, and sole in the North Sea for nearly a decade—warnings that he has come to expect will be largely ignored. It is a challenge he rises to without enthusiasm. Wearing a lugubrious expression, Horwood shows me a corridor lined with photographs of his predecessors. Like him, nearly all are careful-looking men, government scientists ready with a cautiously chosen word for the minister's ear, presented in the detached but pliant manner of the British civil service. This implies that the minister can do what he likes with the information, but what he is hearing is the considered opinion of the Lowestoft Laboratory. One figure stands out in the portrait gallery. It is the only man—they are all men—who is not wearing a jacket and tie and sitting at a desk. He is Michael Graham, director of fisheries research from 1945 to 1958. He is standing on the deck of a trawler dressed in a fisherman's sweater and apron, pipe in mouth. He is busy scaling a cod.

Michael Graham was a scientist whose contribution to the theory of fishing—which chiefly involves describing overfishing—was large and timely. During his time as director of the Lowestoft Laboratory, it achieved world leadership in the understanding of sea fish populations. He was an idiosyncratic employer, ordering that young scientists, or naturalists, as he insisted they be called, go to sea.

To a later generation, trying to explain why the North Sea and other seas declined and their fishing ports collapsed, Graham is a seminal figure, both for his influence on the golden age of fisheries science and for his restless advocacy of "rational" fishing and the conservation of biological systems on which humans depend for food. The starting point for anyone trying to understand his views and ideas on fishing is *The Fish Gate* (1943). In this slim book, in the muscular English of wartime, Graham explains why the idea of free fishing failed in his lifetime and why that failure was compounded by the scientific errors and evasions of previous generations.

When Graham started work at the Ministry of Agriculture and Fisheries' Lowestoft Laboratory in 1920 it had been a demonstrated fact for at least thirty-five years that industrial trawling can cause overfishing. In Britain, where fishing was first industrialized with steam trawlers working out of Hull, Grimsby, and Aberdeen around the 1860s, it was the fishermen, not scientists, who first pointed out that fish populations in parts of the North Sea were being systematically wiped out. The absence of fish in their old grounds required them to steam further out to sea to find new ones. The scientific establishment, in the shape of Thomas Huxley, met this observation with contempt. Huxley chaired a commission in the late 1860s looking at whether fishermen had grounds for concern. The view the commission formed than was still being expressed by Huxley nearly twenty years later, in 1883:

> I believe that it may be affirmed with confidence that, in relation to our present modes of fishing, a number of the most important sea fisheries, such as the cod fishery, the herring fishery, and the mackerel fishery, are inexhaustible. And I base this conviction on two grounds, first, that the multitude of these fishes is so inconceivably great that the number we catch is relatively insignificant: and secondly, that the multitude of the destructive agencies at work upon them is so prodigious, that the destruction effected by the fishermen cannot sensibly increase the death-rate.

Another parliamentary inquiry of which an ailing Huxley was a member reversed those conclusions within the decade.

The decline in the fisheries progressed at an alarming rate over the next twenty-five years and would have continued to do so were it not for the First World War, when most trawlers were either requisitioned for other tasks or unable to go to sea because of mines or hostile shipping. This gave fish stocks a chance to recover. During the war, with little fishing possible, fish prices more than doubled. In 1914, the average daily landing from a trawler in the North Sea was 1,575 pounds. When fishing resumed in 1919, the average daily landing was 3,392 pounds. But within five years the boom had turned to bust. Fishermen who had invested in new boats and gear were catching too much. Prices collapsed, making them fish all the harder. Trawlermen had to sell boxes of 3-foot cod for fish meal. Then stocks became depleted, too. Parliamentary inquiries expressed increasing concern about North Sea fish stocks throughout the 1920s and 1930s.

Michael Graham's given task on joining the Lowestoft Laboratory was to study the North Sea cod fishery. He continued, with breaks for field studies in Africa and Canada, throughout the 1920s and 1930s, describing the cod's life cycle and spawning grounds and showing the age composition of the fishery through a laborious process of scale reading. In a 1935 paper Graham conclusively showed that the stock was overfished. His book *The Fish Gate* reminds us that by 1939 the North Sea was exhausted and many fishermen unemployed—the inevitable result of free fishing. Graham distilled the observations of twenty years at sea into his "Great Law of Fishing": fisheries that are unlimited become unprofitable. Graham said this was a law demonstrable by experience, not scientific theory, an interesting distinction. As long as the effort is free and urged on by competition, fisheries will eventually fail. The Great Law, he pointed out, was fruitful only in its converse form: that limiting the effort will restore profit to a fishery. His next objective was to find a way of showing exactly how much to limit fisheries to provide the optimal catch.

When Graham returned to Lowestoft after the Second World War—which he had spent as a scientific adviser to the air force, appraising the efficiency of new aircraft and equipment—it was as director. Sidney Holt, one of the young mathematicians Graham hired straight from college, remembers it was a time of general optimism and determination that the postwar world would not repeat the mistakes of the past. Half the fishing vessels in Europe had been sunk, and while there were going to be new vessels, Graham believed Europe could avoid building up an excessive fleet, as it had done after the First World War, if it had a clear way of calculating the size of stocks and what fishing pressure would do to them. The job of a scientist was to improve matters, Graham told Holt in his first week at Lowestoft in 1947. The task was urgent: fish stocks had been given a six-year breather that had enabled them to recover from overfishing before the war. One of Holt's first tasks was to look into whether this suspected recovery had actually occurred, which it had. After several years' hard work, he and his colleague Ray Beverton produced their mathematical model of how plaice and haddock populations behave under fishing pressure. Their findings were eventually published in 1957 but long before that began to be internationally influential.

Beverton and Holt had produced the means by which single fish populations could be managed reasonably successfully— provided scientists fed the right data into the models and made appropriate assumptions about a range of things, including the natural and human-induced mortality of fish and provided politicians acted on these recommendations and limited fishing accordingly. These provisos became increasingly important as time went on.

Graham believed that there must be an optimal point some-where in the relationship between a sustainable population size and fishing intensity. Other scientists, including M.B. Schaefer in the United States and W.E. Ricker in Canada, came to define this point later in the 1950s as maximum sustainable yield (MSY). There were some who, unlike Graham, believed that this was an objective to strive toward. Sidney Holt says this characterized the difference between European and American thought during the 1950s. Europeans such as Graham believed that fisheries should be managed for long-term stability and to preserve jobs. Scientists in the United States, mindful that their constitution required any fishery to be open-access, needed to make room for expansion. This is where trouble began.

The pursuit of maximum sustainable yield encouraged fish-ermen to drive down the original population to a lower level—taking up to half of the total spawning stock every year—in the belief that this would boost the productivity of the population. A smaller population, the theory went, would grow faster and reproduce earlier because it had a more plentiful food supply. The problem was simply that the MSY concept tested the Bev-erton-Holt single-stock model to destruction. To decide what level of catches approached the magic figure of MSY there was little room for error. Scientists needed accurate figures for fish-ing mortality (i.e., fishermen must not cheat by misreporting catches or discards) and natural mortality. There were also the dangers of missing or misinterpreting environmental forces or unforeseen predator effects that existed outside their models. Complete accuracy is, of course, impossible, but without it catches will be allowed to rise too high, and the result will be overfishing. A near-definitive demolition of MSY as a concept was written in 1977 by the Canadian biologist Peter Larkin. His short poem on the subject is better known:

Here lies the concept, MSY.
It advocated yields too high,
And didn't spell out how to slice the pie.
We bury it with best of wishes,
Especially on behalf of fishes.
We don't know yet what will take its place,
But hope it's as good for the human race.

Lay people may find it strange that the concept of fishing to provide maximum sustained yield, despite being comprehensively discredited, remains the unaltered objective of several international conventions, such as the one governing tuna stocks in the Atlantic, which was negotiated in the 1960s. Even more strangely, it was included in the part of the declaration of the 2002 Johannesburg summit on sustainable development that applies to fisheries. This said that to achieve sustainable fisheries the following actions were required "at all levels": "Maintain or restore stocks to levels that can produce the maximum sustainable yield with the aim of achieving these goals for depleted stocks on an urgent basis and *where possible* [italics added] no later than 2015."

I was at the Johannesburg summit and remember noticing a flurry of activity until that "where possible" was inserted in the plan for the next two decades. Later on, at a social gathering, the assembled diplomats of the world congratulated themselves on a job well done. Some of us reflected that the self-congratulation was because they now didn't have to do anything at all.

Berlin. A leafier, pleasanter city than I remembered. I ran into Sidney Holt at a meeting of the International Whaling Commission, which he still attends every year, though now in his seventies. We arranged to meet up for a leisurely breakfast around 8 a.m. in our vast, impersonal hotel. We were still talking when the waiters were clearing away the remains two and a half hours later. Holt said he had been looking again at the work of his early years and at the disaster (his word) that had happened to the fish of the North Sea in his absence. Here, it occurred to me, was the epilogue to the golden era of fisheries science being expounded in front of me, as Holt urgently drew dome-shaped yield curves on paper napkins.

Where did he think things went wrong? Thereby hangs a personal tale. Beverton stayed at Lowestoft until the 1970s, when he went off to run the Natural Environment Research Council.

Holt, more of a radical figure, who had been a firebrand communist while in college during the war, went to work for the UN Food and Agriculture Organization in Rome in 1953. In the 1970s he began to work for environmental groups and individuals, lending his mathematical modeling skills to the battle to stop whaling. Holt told me that he and Beverton differed on many things and didn't see each other very often. When they did meet, they argued about where the blame should lie for the increasingly parlous state of fisheries around the world. Beverton argued that overfishing had much to do with authorities failing to act appropriately and in good time on scientific advice. Holt argued—as do contemporary North American scientists such as Daniel Pauly—that aspects of the models routinely used by biologists actually induce overfishing. Just before Beverton's death in 1995, he and Holt eventually agreed that it was a bit of both. In a lecture he sent to a conference in Vancouver that year, being too ill to deliver it himself, Beverton noted: "There is a strong inverse association between the growth of fisheries science . . . and the effectiveness with which it is applied." Kindly, he concluded, "Events of recent years raise in my mind the lingering suspicion that Sidney [Holt] might have been right to the extent that we have been trying to be too clever and have failed to see the wood for the trees."

A prime example of this, says Holt, was the assumption made by scientists for more than thirty years, supposedly based on his and Beverton's work, that the number of juveniles reaching fishable size each year has nothing to do with the number of parents in the sea.

This would obviously be untrue for mammals such as dogs, cats, humans and whales. It would even be untrue for marine creatures, such as sharks, that have few young, so the number of parents is undoubtedly the most important factor in how many offspring fish produce. Fish such as cod, however, can produce upward of 7 million eggs, of which only a tiny fraction survives. For many years it was therefore argued that there would always be enough adult fish to produce the usual number of young, and the main influence on the numbers of juvenile fish were environmental factors, such as temperature and predation. This argument was used to justify higher catches than might otherwise be the case. I remember being subjected to patronizing expositions of this counterintuitive—and, to an informed layman, unbelievable—argument by senior scientists, officials, and even politicians. All brushed aside the layman's observation that logic dictated that the population must at some time reach a level where the number of parents mattered. So what was the level that the number of parents did begin to matter? Was it a low number or a high one? This, as it happens, is precisely Holt's point. He and Beverton found no detectable relationship between the numbers of plaice of catchable size recruited annually to the North Sea plaice stock in the late 1940s and the number of parent fish. But the North Sea plaice was at that time a large population that had just spent six years recovering from overfishing. He and Beverton knew then, he says, that this observation wouldn't always hold. "Even in the 1950s, we knew it didn't and this could cause problems," he said. In the event, it led to "progressively more and more dangerously optimistic quotas."

Another danger with population models is that they usually assume that a small fish population will be more productive than a large one. In nature, however, depleted populations do not behave like healthy ones. They are less productive. This phenomenon is known by a variety of names—depensation, the depopulation effect, or the Allee effect, after the ecologist W.C. Allee, who described it in insects during the 1930s. The Allee effect claimed the passenger pigeon, a creature now extinct, but once so numerous that it used to darken the skies over North America. The same effect kicked in with the northern cod on the Grand Banks of Newfoundland, and is in danger of rearing its head with cod in the North Sea. Alarmingly, the effect is not distinguishable from simple fishing pressure until it is much too late to do anything about it.

The dangerous dogma that the number of parents did not matter meant that scientists were more inclined to blame the failure of stocks to regenerate on environmental variability. There is continual, dynamic environmental variability in the sea, not all of which is regular, like the warm current pulse in the Pacific known as El Niño. This phenomenon was thought to be a factor in the 1970s collapse of the Peruvian anchoveta, the largest fish stock in the world, and greenhouse warming may well be responsible for making the North Sea a less favorable place for cod today. But in these examples there is, or was, massive fishing pressure. Without fishing, which unquestionably kills millions of potential breeding adults, the population would have stood a far greater chance of adapting to environmental change.

Holt told me that rather than aiming high to stimulate fish populations to grow, we should be aiming low, for a catch that is well within the biological limitations of the species, the possible fluctuations of its environment, and possible errors in estimation. That almost certainly implies lower catches than currently permitted in any commercial fishery in the world today. If one were to subject the world's fisheries to the revised management procedure devised for controlling catches of whales, most species of which are now too rare for it to be used, most fisheries would be closed overnight.

Why some overexploited stocks fail to recover while others such as the anchoveta come back is not well understood. Jeff Hutchings, a marine biologist at Dalhousie University in Canada, has analyzed ninety stocks worldwide, many of which have experienced massive declines due to overfishing. Most, with the exception of the fast-growing Atlantic herring, showed little signs of recovery fifteen years after their collapse. Hutchings told *Nature:* "The life history of species matters. Small, early-maturing, mid-water species like herring might recover faster than late-maturing, bottom-living species such as cod." Hutchings's view was that if stocks were not recoverable, a far greater degree of precaution was needed in setting catch limits in the first place. "If there is not much we can do after the damage is done, then it is an even stronger case that we should not let fish stocks fall below safe levels." Estimates of sustainable catches have become much more conservative since the 1970s, and more so since the Grand Banks disaster, but almost nowhere, with the arguable exception of Iceland and for a number of pelagic species, have catch limits been set low enough for populations to rebuild.

The Grand Banks is the textbook case of failure in fisheries science. An army of scientists in one of the world's wealthiest and most advanced nations managed to destroy one of the richest fisheries in the world, while convincing themselves for a decade that they were doing no such thing. The Newfoundland cod collapse was the nightmare that shook the world out of its complacent assumption that the sea's resources were renewable and being managed in an enlightened manner.

It is not possible to say that with better science Canada would have avoided disaster, because there is no way of predicting what pigheaded decisions politicians would have made if scientists had noticed earlier that the cod was in trouble. Politicians in Canada have a long record of using any kind of scientific uncertainty to argue for higher catches. What one can say is that according to the models scientists were using in the early and mid-1980s, the Grand Banks should still be full of cod. The great advantage of hindsight is that one becomes aware of the dissident voices prophesying disaster that were ignored by the bureaucratic machine that was Canada's Department of Fisheries and Oceans (DFO), and by its elite and secretive bunch of scientists responsible for assessing stocks. As far as one can see—and this area is fraught with scientific papers trying to write and rewrite history to excuse some and blame others— a succession of mistakes was made, which led to stock assessments being grossly inaccurate beginning in the late 1970s. All the mistakes involved assumptions fed into population models, which proved the old computing adage: garbage in, garbage out.

The first glaring error was made in the late 1970s and early 1980s, when Canada declared jurisdiction over 200 miles of the Grand Banks and ejected trawlers from the Soviet Union, Poland, Britain, Spain, and other nations that had hammered cod stocks there to a tenth of their level in the 1960s. There was a political opportunity, between 1978 and 1983, when scientists could have set a more cautious course because the Canadian fleet did not have the capacity at that time to overfish the cod. But stock assessments of cod in the late 1970s failed to reveal the extent of destruction to the spawning stock, which was eventually revealed by scientists going back over the same figures a decade later.

What happened was that Canada did not have credible data for some of the areas within its new fishing zone—because Germany was the last nation to have conducted surveys there when the Grand Banks were international waters. As a result, commercial catch per unit effort data were used to "adjust" or "tune" models that were supposed to be based on random trawl surveys. The assumption that the commercial catch data conveyed an accurate picture of stocks across the seabed, rather than on the grounds where fishermen went to fish, turned out to be catastrophically mistaken. Where commercial catch data showed the stock had declined by 70 percent, the stock had actually fallen by 90 percent since 1962.

Convinced that cod stocks would now recover quickly to pre-1960s levels, the provincial government of Newfoundland and the government of Canada called for long-term catch predictions based on these erroneous assessments. Scientists duly predicted catches of 400,000 tons a year by 1990. These predictions now look incredible. Stocks never rose enough to allow catch limits higher than 260,000 tons. Initially the cod did stage a recovery, though never to the levels predicted. By the time it was discovered, much later, that commercial catch data and trawler surveys were showing different trends, it was too late.

Sidney Holt identified a number of now-classic errors. The importance of a large, healthy spawning stock was ignored. The overfished population did not become productive; it turned out to be less productive than the previous larger one. By the early 1980s, on the basis of rosy long-term forecasts, heavy subsidies were pouring into the province to build Canada's own trawler fleet. Carl Walters of the University of British Columbia and Jean-Jacques Maguire of the DFO in Quebec published a paper in 1996 that concluded: "Even if scientists had suddenly reversed their conclusions and called loudly and publicly for harvest restraints, by about 1982 an institutional juggernaut had been set in motion that no political decision-maker would have dared try to stop until it was too late. The window of opportunity had closed." Astonishingly, Canada's scientists continued to think that they were setting annual catch limits at 16 percent of the fish, which in theory would allow stocks to increase rapidly. Later analysis would show that fishermen were catching more like 60 percent of the adult fish each year. Warning signs began to appear: smaller fish, an indication that they stood less chance of surviving; the dragger (trawler) fleet was fishing a smaller and smaller area of ocean; inshore fishermen complained, as they had since the early 1980s, that catches were going down. To compound the muddle, the technology on board draggers had increased as a result of subsidies, thereby increasing the catch per unit effort—one of the ways scientists measured the stock. But the Department of Fisheries and

Oceans had no way of revising its estimates to take account of technological creep.

Every autumn the DFO research vessel steered its random course across the banks, counting how many fish it caught. By 1989 it was showing large areas of empty ocean. But fishermen said there were still plenty of fish because their fish finders were detecting hot spots where the dwindling shoals of cod were congregating. Schools of cod or haddock huddle together when they are depleted. Other fish, such as hake, do not.

The scientists couldn't decide what to do. The whole theoretical basis on which they had built their assessments was crumbling. They advised a catch limit of 125,000 tons, less than half the 266,000 tons of 1988. The fisheries minister of the time refused to agree to it for fear of upsetting fishermen. He set the quota at 235,000 tons. Lesley Harris, a former president of Memorial University in St. John's, Newfoundland, said that the DFO should have insisted. "But scientists being scientists, they weren't prepared to make absolute statements about anything."

Many remember Jake Rice, chief of research at the DFO in St. John's, cheerfully going on television before the collapse, saying that lots of young cod were coming through, several year-classes of them. These young cod never made it. Millions were caught and discarded in the last two years of the fishery by fishermen who refused to believe how serious things were. By June 1992 the DFO realized that there were no cod left old enough to spawn, and by then even the fishermen were showing concern. A year-long moratorium was declared and extended indefinitely in 1993.

The failure to be truly open with data and to go public about uncertainty is perhaps the greatest lesson for science from the Grand Banks disaster. With more transparency, the errors would have been detected earlier. Ransom Myers, then at the DFO, says he was not allowed access to the data used to calculate stock assessments, as it was the preserve of a charmed circle he calls "the tribe." Members of this tribe, he believes, were unduly concerned with protecting the reputations of the assessors and their political masters, and this got in the way of accuracy. Why does he think a scientific team, supposedly paid to reward its independence and expertise, failed to spot such huge errors for so long? "Every choice," he told me, "was about 'What is going to get me rewarded?' Positive news makes people happy. Bad news is not always appreciated."

Professor John Shepherd of Southampton University in England, a former deputy director of the Lowestoft Laboratory and one of the small group of fisheries scientists to emerge from the British government machine with integrity, once put it better than anyone else: "We will always have a problem until we recognize that a scientist's first duty is to the truth. His second duty is to the public interest and his third duty is to the minister."

What else does the Grand Banks crash tell us? Could such huge scientific errors happen again? The answer, according to Carl Walters and J.J. Maguire in 1996, is yes, because systematic overestimation of stock size and inadequate attention to the spawning stock are common problems, difficult to expunge from single-stock models. Maguire became head of research at ICES and helped it to clean up its act by promoting transparency, cracking down on false assumptions in computer models and promoting a greater degree of precaution in deciding catch limits. Since then ICES has consistently recommended a total ban on catching cod in the North Sea. But many other scientific bodies, notably those responsible for tuna, are suspected by other scientists of still coming up with gung-ho stock assessments that bear the hallmarks of wishful thinking.

One positive development since the Grand Banks disaster is the number of fisheries scientists attracting funding from conservation bodies. Diversity of funding is good, for it overcomes the traditional scientists' problem that whatever the data say, the paymaster—generally a government—gets to call the shots when the summary is written. More open diversity of opinion makes fisheries scientists have to defend the catch limits they recommend not just to self-interested fishermen, who might wish them to be higher, but in the arena of world opinion against knowledgeable people from related disciplines, who may well think the quotas are far too high.

So what are we to conclude about Michael Graham and his colleagues in the 1940s and 1950s, who constructed theoretical tools that they hoped would enable the sea to be managed in a rational, benign way? Did they try as hard as they might, or were their efforts thwarted? Cynics would say that the science of counting fish merely served to stretch out the period between boom and bust for forty years, from the late 1950s to the late 1990s—in other words, scientists simply measured decline without actually managing to reverse it when things got tough. Actually, the only accurate way of judging scientists—from Europe or anywhere else—is to ask whether, if politicians had taken their advice, the fish stocks of the northeast Atlantic would now be in a better state. The answer, unequivocally, is that there would be more fish.

On the eastern side of the Atlantic, scientists have made many errors, but the dire state of Europe's sea derives overwhelmingly from poor governance. If you look at the declining line of fish catches on the graph over the past fifteen years,

ICES got the trend right. You might rightly criticize it for failing at any time to recommend that quotas be slashed heavily to allow stocks to rebuild. To an outsider there does seem to be a general complicity between the EU's scientists and politicians that they will manage stocks for stability at whatever level they slump to next—some call it bumping along the bottom. Nevertheless, the best judgment has to be that if ICES's advice had been taken for the North Sea's white fish stocks—cod, haddock, and plaice—in the past decade, as it was with the herring, the sea's fish populations would be in less of a desperate state.

The failure to act on what *is* known was just as bad after the war. When Michael Graham arrived back at Lowestoft in 1945, a considerable amount of thinking had already been done. In 1941 his mentor, Dr. E.S. Russell, chaired a committee of scientists looking at ways to ensure that the fishing grounds of the North Sea would not be overfished after the war if the Allies won it. The committee made a number of astonishingly modern-sounding proposals. First was that there should be regulation of the number of days each vessel spent at sea. Another was that the tonnage of the fleet after the armistice should be set at 70 percent of the fleet in 1938. A third was that there should be a minimum mesh size, but this should be larger than that agreed at an overfishing conference in March 1937.

Then a process of political dilution was applied to what the proposers had intended as a coherent package. The Admiralty and the Department of War Transport objected to the proposal to cut tonnage for military reasons. A war cabinet committee ruled in March 1945 that Russell's recommendations should be applied only to North Sea fisheries, not those in the North Atlantic and Barents Sea, as the evidence of overfishing there *"was not absolutely conclusive"* (italics added) and because negotiations would involve several countries, including the Soviet Union, so were unlikely to succeed. The recommendations on the North Sea, however, were accepted. The Foreign Office was instructed to convene an international conference on overfishing as soon as possible.

What happened at that conference in March 1946 is worth repeating because it so closely resembles what has gone on in Europe ever since. The United Kingdom pressed for the restriction of fleet tonnages. This was not accepted because it did not suit the postwar regeneration plans of other countries. The conference recommended instead that the fleets be kept at their present size, or that of 1938, whichever was greater. Even this measure was not acceptable to Denmark, Norway, and Sweden and was accepted by Spain and Iceland only on the understanding that it did not interfere with their plans for building new vessels. Ideas for a total allowable catch, closed seasons, and comprehensive fishing trust with exclusive rights over certain fisheries all foundered. The only agreement from the conference was for a convention to regulate mesh sizes. It took until 1953 for all signatories to ratify the convention, after which a permanent commission was set up in London to run it, with a secretariat provided by the ministry. This commission was unable to resolve disputes, such as those that broke out between the United Kingdom, with its large, long-distance fleet, and Norway, Iceland, and the Faroe Islands. The failure of these postwar arrangements led to the "cod wars" of the 1960s and 1970s. Why did these arrangements fall apart? It was the same old story: fishermen vote, fish don't.

Thus was the great opportunity to manage Europe's fish stocks for future generations squandered. It will take the equivalent of a world war to bring the fish back again.

AFTER THE GOLD RUSH

Bonavista, Newfoundland. Fall. Nothing prepares you for the beauty. When I turned off the highway on to the Bonavista peninsula the road soon began to wind around the rocky coast through fishing settlements, known as outports, and the setting sun burst out of a stormy sky. The squat orange maples and yellow birch glowed brilliantly against stunted fir and spruce. As I looked inland, vistas appeared toward the hills, over peat bogs and what in Scotland would be called black lochs. Seaward, along the rocky promontories and bays, the trees were few, but the clapboard houses looked welcoming, and one inhabitant took great pains to show me the way. The wooden stages, or flakes, where cod was once dried looked like relics of an earlier, more prosperous age. When you think of it, that is exactly what they are—remnants of a "gold rush" that lasted nearly five hundred years.

Nothing quite prepares you for the distances, either, in a province the size of England that was, until 1949, a separate country and has its own time zone, half an hour different from the Canadian mainland. The plane landed in St. John's at 1:30 p.m. in rain, low clouds, and 70 mph wind, and I was drenched before I reached the rental car. It was 200 miles to Bonavista by road. Having been warned not to drive after dark because of moose on the roads, I drove north up the main highway as fast as I dared. The radio was tuned to a station with a 1980s playlist that included Tom Petty and Fleetwood Mac, varied occasionally with more contemporary singers. This produced a sensation of drifting in and out of the past that I was not to shake off until I left Newfoundland. It was dark when I arrived in Bonavista, and I was probably as grateful to see it as John Cabot was when he sighted the windswept cape from the sea in June 1497, exclaiming, *"Buona vista!"* (What a fine sight).

My sense of being in a time warp was heightened further when I read the tourist information about Bonavista (compiled in the early 1990s) provided in Abbotts B&B on the road out of town toward the cape. Typed sheets of paper in plastic folders told me that the town is windy all year round but does not have really cold weather or hot summer temperatures because it is at the end of a peninsula. The spring is short. Icebergs drift down the coast well into May, when the weather can be miserable. Only the brave swim in summer. Bakeapples, otherwise known as cloudberries, grow in the marsh. They are noted for their taste. So, too, are partridgeberries, which contain a natural preservative that means you can keep a pot of them in water all winter to bake in muffins. Youths amuse themselves by persecuting frogs in the marsh and driving around on quad bikes. The tourist information lists local issues and concerns: whether the Arctic ice will remain inshore into the fishing months, for if it does, it will lower incomes; fish prices; the availability of fish species; the increase in whale and seal numbers. Under the heading "Daily Life in the Community" some thoughtful Abbott had recorded: "The fishery is Bonavista's main industry.

Cod drew the Europeans here in the 1490s and keeps us here in the 1990s. Without the fisher, Bonavista could not survive. IN COD WE TRUST."

Nothing had been added to this passage since 1992, the year the Canadian government closed the cod fishery. Perhaps the author thought, as did the Canadian government of the day, that the crisis would be over in a couple of years. Perhaps no one in the Abbott household quite knew how to come to terms with the end of a period of dependence on the codfish that began nearly five hundred years earlier with the arrival of John Cabot. Born in Venice as Giovanni Caboto, his real name was anglicized to stress the official nature of his mission. He had been commissioned by Henry VII to find a western passage to Asia. What he actually discovered in 1497 was a northern landfall on the American continent from which all Britain's subsequent claims in the New World derived, and cod in such quantities that, he claimed, all you had to do was lower a basket from the boat to catch them. At the time of Cabot, Canadian scientists estimate that the amount of spawning cod off the Canadian coast amounted to more than 4.4 million tons. By 2003 there were about 55,000 tons.

Remote places often give the warmest welcomes, and Bonavista prides itself on the friendliness of its people. I arrived in the dark and the rain on a Sunday evening to find my main contact had gone home. Next door was the town's café, a place that was open but wasn't quite finished yet. Its proprietor, Harvey Templeman, saw a tired, jet-lagged figure with no Canadian money. He pushed across a cup of free coffee and busied himself with the phone directory until he had found the home number of the person I wanted from the mayor's office.

I came to Newfoundland to try to understand for myself why

the cod collapse happened, how it affected the place, and what hope there was for the future. Some 44,000 people in fishing and processing are said to have been put out of work in 1992—an enormous number. One wonders what exactly they were all doing. In the early 1990s there were 705 jobs in Bonavista directly provided by the fishery, in catching and processing. The next day I asked Betty Fitzgerald, the mayor, what happened to them. She told me that the town had lost 700 people since 1992, though not all were fishermen. Since then, inward migration had started again, with people moving up from Ontario and the United States. They had discovered the place through tourism, which hardly existed before.

The town has survived in the absence of cod by placing its trust in its heritage. Interest began with the five hundredth anniversary of Cabot's landing. A replica of his vessel, the *Matthew*, was finished in 1998 and now overwinters in a specially made boathouse in the harbor. Heritage trails were created along the Bonavista peninsula and in other regions of Newfoundland. Since then, an increasing flow of tourists has arrived to see John Cabot's statue and the lighthouse on the end of the cape, and mostly just to drive around. Tourism hasn't quite made up for fishing yet, but Betty is working on it. She has successfully sought heritage grants for historic buildings, of which Bonavista has more than a thousand—more than any other settlement in Atlantic Canada. The oldest is a house built in 1811 by Alexander Straithie, a tradesman from Renfrew, Scotland. Fixing up old buildings provided employment for enough weeks for fishermen to be entitled to employment insurance for the rest. But Betty Fitzgerald still has five hundred people seeking two hundred jobs. She's getting a theater started and trying to get a slate mine opened again. She wishes she could get an old sealing plant reopened. Bonavista is trying all the things that Cape Cod fishing communities did when they ran out of fish.

In terms of tourism, Bonavista has a lot of advantages compared with other parts of Newfoundland. After Cape Spear, the most easterly part of the continent, just south of St. John's, it is the most obvious historical site that any visitor might want to see. Only now is the town beginning to wake up to its potential, so I was surprised when Betty described the cod fishery as the backbone of the community.

Her three sons still work in fishing or fish processing, although they have struggled over the past decade. There were still plenty of fish in the bay, she told me, so she couldn't understand why the inshore fishermen—who fish within the 12-mile limit—couldn't use their handlines and traps to catch it. That was the strongly held local view, explained to me patiently by Larry Tremblett, an inshore fisherman with twenty-three years at sea, who now faces a fine of $440 (in U.S. dollars) if he catches a single cod.

The bizarre thing is that Bonavista Bay and Trinity Bay are full of cod. Sitting in his 35-foot boat, *High Hopes,* in Bonavista harbor, Larry explained: "That's the way it has been round here for five years. When we put out lumpfish nets, or try to catch blackback [winter flounder], we get cod. We get cod of 50–60 pounds sometimes. Nobody knows what the Department of Fisheries and Oceans is up to. They just don't want us to get at it. They think the inshore fishery is a pain and they are trying to freeze us out." There is bad blood between the inshore fishermen and the DFO, which is based in St. John's and Ottawa. The DFO ignored warnings from the inshore fishermen in the 1980s that their catches were declining and that big draggers, as stern trawlers are called, were overfishing the Grand Banks cod. The inshore fishermen were right, but the DFO scientists said it was too complicated to measure what the inshore fishermen did, so the fishermen's claims were never tested. Now the DFO says that the fish in Bonavista Bay and Trinity Bay are all that remain of what was once the largest cod population in the world. The northern cod once used to spread along the coast of Newfoundland and Labrador and 240 miles or so out on the Grand Banks. (There are actually ten separately spawning stocks of cod off the Canadian coast, but the northern cod was always the largest.) Even though the DFO has almost certainly got it right this time, the inshore fishermen don't believe the DFO on principle. It's probably not such a bad principle, given what they've been through.

The history of the Newfoundland cod fishery looks different from the inshore fishermen's perspective. By the early 1990s Larry Tremblett and Doug Sweetland, for example, saw the population of cod begin to retreat south, until there was none off the coast of Labrador, then none off the north of Newfoundland. Next the inshore fishermen noticed that the fish were getting smaller. The Fishermen's Union, which represents the inshore fishermen, pointed this out to the DFO and called for a ban on dragging the spawning grounds. As Alan Christopher Finlayson has pointed out in his book *Fishing for Truth*, the DFO preferred to work with data from the big companies that ran the trawlers. "They used to look at us as if we didn't know what we were talking about," said Tremblett. "We proved them wrong. If they had started listening to us in the 1980s, we wouldn't have had the problems we have today."

Doug Sweetland is even-handed in dishing out blame. "Everyone contributed to the decline of the cod. The main culprits were the draggers because of high grading." High grading is where fishermen keep catching fish and throwing away the ones they don't want until they have achieved a full hold of premium-size fish. "The plant in Catalina employed 1,500 people. They were getting so much fish in the mid-1980s that they said 26 inches is the smallest fish we want. You know where the rest was going—over the side. That destroyed a million pounds of fish to get 400,000 pounds." Sweetland remembers how quickly the end came. "In the winter of '92 there was good cod. Within three months there was nothing. The offshore fishery had shot itself down." Ironically, it was the inshore fishermen, who caught least, who have suffered most. Of the big companies, FPI is back in business processing shrimp and crab, which it buys from boats that it does not own.

For those fishermen who used to catch cod close inshore, the future looks bleak. The state of the northern cod twelve years into the crisis is now worse than it was in 1992, thanks to another foul-up by the DFO and its political masters. Back in 1992, those scientists who recommended closure of the fishery predicted that the stock would recover to fishable proportions in two years. Working on that assumption, the Canadian cabinet decided to pay fishermen a $3.5 billion "package"—for social security, license surrender, and retraining—that eventually ran out in 1995. Still there was no recovery. The treasury refused to subsidize the fishermen further, as the disaster was at least partly of their own making. Strangely, it made no such decision about the DFO.

As a result, fishermen in rural Newfoundland pressed for a small-scale reopening of the fishery. The government had run out of options. The fishery was reopened, contrary to the advice of many scientists led by Ransom Myers, by then no longer at the DFO. This so-called sentinel fishery was meant to contribute toward scientists' knowledge of stocks by recording the age profile of the cod that was caught. Although it was restricted initially to traps and handlines, it was a commercial fishery. Owing to the number of fishermen participating, it proved hard to monitor catches or cheating. Catches were almost immediately unsustainable, according to Peter Shelton, a DFO scientist, who was responsible for monitoring the result. The sentinel fishery was closed down in spring 2003 after DFO scientists concluded that "serious harm" had been done to stocks. The big difference was that this time the minister and the DFO accepted that there was no chance of things getting better in the near future. The best estimate of a recovery time was fifteen years. Nobody actually knows if the cod will ever recover.

The second foul-up by the DFO has given Bonavista's inshore fishermen very few options. For although they might want to catch other fish, what they do catch when they shoot their nets or traps is cod. Tremblett was fishing for blackback in Bonavista Bay and worried himself sick when he inadvertently caught 3,000 pounds of cod, until officialdom let him off. He now gets by fishing for crab for just a few months of the year. He makes enough in the fourteen-week spring season for crab to qualify for unemployment insurance, around $600 a month, which the government pays for the rest of the year.

I found inshore fisherman Doug Sweetland at home after he had returned from moose hunting in the north. "Got to scrape a living," he told me with a grin. Doug catches crab, lobster, and lumpfish to qualify for his unemployment insurance. Sweetland told me the extraordinary story of the time the remnant northern cod died of cold.

For some unexplained reason, about 22,000 tons of cod, sizeable fish of 8–10 pounds, shoaled during the winter of 2003 in Smith Sound, south of the Bonavista peninsula. The water was really cold, and perhaps because so many cod were gathered together, lots of them were pushed up from the bottom into the coldest water, where ice was forming. The ice blocked the cod's gills and they weren't able to extract oxygen from the water, so they were smothered. Sweetland said: "The scientists couldn't believe how cold the water was. Two million pounds of cod was just floating there, dead. You were allowed to pick it out." It is not the only time this has happened: the first recorded occasion was seven years earlier. There is a plausible link with overfishing: as noted, cod and haddock are known to huddle together when they are depleted.

The ban on catching cod for the foreseeable future has sharpened the fishermen's resentment of seals and foreigners—usually, but not always, in that order. Sweetland calculates that the amount of cod caught by fishermen in the sentinel fishery amounted to 36,000 tons over the past six years. He calculates that hood and harp seals, of which there are 6 million or more on the Newfoundland and Labrador coasts, ate 55,000 tons of fish a year. "The scientists said that didn't affect the stock. Only what we caught affected the stock. Do you believe that?"

Did seals really eat cod, a fish that lives on the bottom many hundreds of feet deep? I asked Sweetland. In reply, he me told the story of a time he was out shooting terns—legal if for personal consumption in Newfoundland. "We saw this seal come up in 100 fathoms [600 feet] of water with a fish in its mouth. The fish was alive. The seal let the fish go and we turned the boat back just to pick it up. It was a cod." The other all-too-real threat to the remaining cod comes from foreign fishermen, who continued to fish the "nose and tail" of the Grand Banks outside

Canadian waters—and possibly *in* Canadian waters at night. Reports by observers show 5,500 tons of cod, Greenland halibut, and other moratorium species still being caught in the area managed by the North Atlantic Fisheries Organization outside the Canadian exclusive economic zone. Betty Fitzgerald said she has been out on the Grand Banks since 1992 and seen the lights of foreign trawlers "like a city at night." It is a common reaction among fishermen everywhere to blame foreigners for overfishing. That does not mean they're wrong.

Some might say that the logical thing, if Bonavista can't catch fish, would be to make a virtue of necessity and declare the bay a marine reserve and make it a tourist attraction. This is not a popular option. Parks Canada, which runs the national parks, tried to impose just such a solution. The bureaucrats in Ottawa miscalculated badly when fishermen found out, as the bill was going through its second reading in Parliament, that the area they had been told was excluded was not. "What the guy was saying and what was happening were two different things." They wouldn't have been able to control the seals, said Sweetland, and when Parks Canada set up a national park on the coast in the 1940s, they burned all the jetties and forced out the fishermen. Faced with implacable local opposition, the marine reserve did not get through. Sweetland has more time for a marine protected area, which a group of lobster fishermen has set up under the DFO, to buffer the lobsters against overfishing in Trinity Bay. Sweetland wants to go back to catching cod. He thinks the stocks in Bonavista Bay would stand it if inshore fishermen like him had a quota of a ton a head.

Before I headed back to St. John's, I dropped in on Harvey Templeman again. He was outside, painting the shop in his overalls. We stood and looked at the town: the harbor, the courthouse, the busy intersection in front of us, where many of the trucks passing were those of semiemployed fishermen. "You do wonder about these fishermen sometimes. They say they are hard up, but they're still driving around in big trucks." Templeman knows what the outside world is like: he spent four years in New York City. He is a businessman, in a pretty laid-back way, and he can see the place changing. "People are discovering that tourists are coming now. They are beginning to understand that it can be a business. There weren't any tourists before 1997."

I told him, because he asked what I'd learned, that I found it an amazing fact that fishermen face large fines for catching cod while there are loads of fish in the bay. His girlfriend said, "It's not amazing. It's outrageous."

Bonavista is dealing with the present. But, like the rest of rural Newfoundland, it hasn't given up on the past. It wants to catch cod, as it has always done.

Back in St. John's the next day, I talked to Alastair O'Rielly, who was busy winding up the Fisheries Association of Newfoundland and Labrador—the organization that represented fish processors and companies that operated the big dragger fleet. Its members were frustrated with the glut of capacity in the processing sector and their lack of success in influencing government, and wanted a new start. He explained that a gulf that no one has yet bridged exists between fishermen in the north, where I was, and scientists.

The only remnants of the northern cod are in Trinity Bay and Bonavista Bay. The people in that area hold the view that the stock is healthy. No one else holds that view.

This body of cod has been analyzed to death. There is a

spawning biomass of 50,000–60,000 tons where once there was 1.5 million. There's nothing left. We have 380 shrimp vessels fishing on the Grand Banks. We have no cod bycatch problem. We would love to have that problem because it would give us some indication of recovery. It doesn't exist. DFO do a survey with a shrimp trawl. It catches everything. They don't find cod out there either. It doesn't exist. There is no indication that recovery has begun or is even possible.

On the other hand, O'Rielly explained the extraordinary upside of the cod crash. Harvesters, as he calls fishermen, have enjoyed a windfall. Thanks to the absence of cod, which used to eat shellfish, there has been such a bloom of shellfish, snow crab, and shrimp that the industry as a whole is making more money than it did in the late 1980s. Total production of seafood has exceeded $880 million in the past five years, while it was only $704 million in the late 1980s. A further advantage is that while the value has gone up, shellfish being worth more than cod, the tonnage has halved, thereby halving the shipping costs. The only people who haven't benefited from this bonanza are the inshore fishermen with limited access to crab and shrimp, which are caught mostly well out on the Grand Banks by the bigger boats. O'Rielly said, "Nature abhors a vacuum. We're very fortunate. It could have filled up with something the world doesn't want." There was a time, he told me, when you would step on a snow crab if it landed on deck because there was no market for it. Again Newfoundland was lucky. All the marketing work in Japan had been done for Alaskan snow crab, which collapsed just as Newfoundland snow crab became available in large numbers.

The ecosystem shift on the Grand Banks, which may or may not be reversible, has made Newfoundland the largest producer in the world of snow crab and the largest producer of cold-water shrimp. Catches of snow crab went up from 17,600 to 75,900 tons in seven years. The result for fishermen equipped to catch crab and shrimp was incomes at levels they would not have thought possible in 1992. There is only one problem. No one, as Bruce Atkinson, regional director of science, oceans, and environment at the DFO in St. John's admits, knows what a sustainable catch of snow crab or shrimp is. What scientists do know is that shellfish populations are volatile. O'Rielly added: "If you want to worry in Technicolor, you'd worry that something happens to the shellfish resource. People haven't given it that much thought."

The DFO has tried to set catch quotas on a precautionary basis. In the case of crabs, only males that have reached terminal molt are taken. In the case of shrimp, about 12 percent of the total biomass is taken each year, half of what is calculated to be sustainable. In theory, the catch could be doubled, but that would be taking more risk than the industry wants. O'Rielly said:

> If we'd had a wish list since the moratorium, it would have been for crab numbers to be up 400–500 percent, shrimp up too, for the Canadian dollar to depreciate, and for our competitors in Alaska to decrease production. And we would have liked to see consumer willingness to go on paying high prices for shellfish. That's what we asked for and that's what we got. We just don't know how long it will last. It makes you nervous. So much is riding on it, not only for the harvesters, but also the communities

where they live. There is not much else in the rural economy. We're it.

You would have thought that some people might actually be happy with the way things have turned out. To my surprise, O'Rielly, like everyone else, wants the cod back. This, I got the feeling, is not only because he thinks a more complicated ecosystem might be more stable. Catching cod just feels right in Newfoundland, whether you are an inshore fisherman or have a university degree. People are happier with what they know and believe is natural. O'Rielly, like so many others, feels that the number of seals around is not natural. Each of the 6 million harp and hood seals, he told me, eats around 1.4 tons of fish a year. "Just as long as we understand this is a choice we are making. The consequence is that twelve years into this crisis it looks as if we have been replaced by seals and whales at the top of the food chain."

The biggest problem Newfoundland has faced yet will come, paradoxically, if the cod does increase. If the cod gets more plentiful, it will eat the shrimp and the small, soft-shelled crabs. As Bruce Atkinson of the DFO in St. John's put it: "You think things are bad now? You're going to have a time when *everything's* shut down." Atkinson knows better than anyone the political pressures that situation can bring in a province where up to eight thousand people still work in fish processing and as many as fourteen thousand in what the industry calls harvesting. He observed ruefully of 1995: "There was no longer-term plan in the 1990s. At the first sign of a few fish, everyone wanted to get at them."

That's the problem. So what makes him so sure that the same cycle is not going to happen all over again? Is there now a long-term plan? Two things have certainly changed since 1995. One is the government's undertaking to favor the precautionary approach, which says that the fish population must be preserved above the level at which recruitment is impaired; the other is the 1996 Endangered Species Act, which does not allow fishing of endangered stocks. But what size must the cod stock return to, assuming it ever returns, for fishing to be allowed again? How many fishermen should be allowed to fish it? Nothing appears to have been decided. You wonder if any lessons have really been learned by the DFO, by politicians, or by the fishermen themselves. Already there is pressure to reopen the cod fishery in the northern part of the Gulf of St. Lawrence. Will the DFO get it wrong a third time? While individual scientists undoubtedly strive for objectivity, the DFO policy-making machine is well attuned to Newfoundland's style of politics.

Although a visitor to Newfoundland, I realized that I had now begun to get the measure of its cozy, handout-ridden culture. Political success is equated with getting more money out of Ottawa and almost never with conserving fish. The time has come to say that Canada's system of unemployment insurance amounts to a massive subsidy to fishermen to stay where there are. The subsidy ensures that there is a fully equipped fleet ready to go fishing the moment there is even a small amount of fish to catch. Ransom Myers puts as his number one reason for the cod collapse the Canadian government's introduction of unemployment insurance in the 1960s. "The only equilibrium in a subsidized system is zero fish," he told me. "The system is set up to fail—necessarily."

Myers studied the fishery from records made in the seventeenth and eighteenth centuries. He found that when catches were good, people moved into Newfoundland; when catches were bad, people moved out. Four centuries ago, each boat could catch and process 10 tons of fish, salting and drying it, which was a very labor-intensive process. Four centuries on, many boats were catching exactly the same amount. The efficiency had not changed at all, because of subsidy. With unemployment insurance, it is not inconceivable that a couple can make $60,000 a year, spending fourteen weeks catching or processing crab or other fish and the rest of the time improving their lives. For fishermen, all business expenditure is allowable against tax. The time they don't spend fishing can be spent fixing their home, shooting moose for food, swapping labor with their friends, or chopping wood. People enjoy hunting and gathering, provided their basic income is ensured. An enviable life in many ways—but why, you may ask, should the taxpayer pay for it?

"It's totally crazy," acknowledged Myers.

Too many people in the fishery create tremendous political pressure for rules to be bent and quotas to be increased. Myers, a former DFO man, says you won't hear a peep of criticism of this state benefit system coming from the DFO's big, modern building in St John's. "For an academic in Newfoundland to say anything against employment insurance is totally unthinkable. It's not even discussed."

Why do people, all over the world, think that there is something so valuable in coastal settlements that mean they must be subsidized? Why are people willing to subsidize fishing to an extent they might not contemplate with other industries, such as coal or even farming? Myers thinks it is because the hunting-gathering lifestyle appeals to something in all of us. That may be true, but the justification for it, post–cod crash, is definitely past its sell-by date.

Myers is not entirely right about the DFO failing to grasp the nettle of subsidies. A rare example of it acknowledging the pervasive influence of subsidy comes in a paper by Jake Rice, Peter Shelton, and three other scientists, looking back at the cod crash and forward to the future. It sounds increasingly like Myers himself. In it the scientists point out that the auditor general of Canada investigated the $3.5 billion Atlantic Fisheries Adjustment Package, the social support scheme that existed until 1995. The package paid a minimum income support of $350 a month to fishers and plant workers who had lost their jobs. The package paid double this to workers who agreed to retrain and learn skills, such as computing, that were readily transferable back into fishing. The auditor general found that most of those who had been retrained said they would return to fishing as their primary job just as soon as it was possible to do so. Profits from the new crab and shrimp fisheries tended to be reinvested in technologically sophisticated vessels equipped to take part in a range of fisheries, including cod, if there was any to catch. The result was that after spending $3.5 billion to "adjust" the Atlantic cod fishery, the effective fishing capacity was 160 per cent of what it had been in the early 1990s.

When fishing resumed, Rice and his fellow scientists continue, the imperative resulting from so many fishermen needing fourteen weeks' income so that they could claim unemployment insurance made the participation of the maximum number of people, rather than profitability, the governing factor in the allocating quota. A large number of very low quotas were therefore doled out, which proved hard to monitor and enforce. This means a lot of cheating went on. The officials at the DFO do seem at last to have recognized that the costs of excessive optimism are greater than the costs of being unduly pessimistic.

They say the prospects of a cod recovery in the next few years are very unlikely. Even small fisheries could annihilate any gains and prevent the stocks from returning to their historical biomass and yields. Although the wily DFO men could just be distancing themselves from blame when Canada gets it wrong all over again, I think their summing up has the ring of truth. Rice and his co-authors conclude: "At least in Canada, fishing is a culture, not just an economic activity. People will come back to the fishery when it reopens, in large numbers and with new skills and high expectations." They say that unless excess capacity (which means fishermen and fishing vessels) is *permanently* removed from the fishery early in the recovery, the expenditure on science and management needed to ensure the stock recovers will turn out to have been pointless.

No one should underestimate the political difficulties of stopping fishermen from fishing in a province where so many of them vote. Yet somehow it must be done, or the cycle will go on repeating itself. The cod will recover—just, maybe—and will be fished out all over again. Until someone has the courage to strike out in a new direction, Newfoundland will just be a desperately sad place, making the same mistakes over and over again without ever appearing to learn from them.

SUBSIDIES

A subsidy is a sum of money dished out by the government that allows a commercial venture to go on doing something beyond the point at which it would otherwise have gone bust or been forced to do something else. Despite the increasingly alarming state of the world's fish stocks, there are plenty of countries still doling out subsidies to fishermen that will ultimately make things worse for fish—and therefore eventually for fishermen.

Initially, the taxpayer might benefit from plentiful supplies of cheap fish, but he or she will end up paying twice—once to pay fishermen to fish and again when the price of fish rises because stocks have been overfished. The unfortunate taxpayer may even have to pay a third time, as in Canada, when the fishermen have to be paid *not* to fish because stocks have disappeared. Canada, however, is by no means the worst in the subsidies racket.

Globally, the amount of subsidy in fishing is very large—as much as a staggering $50 billion a year, according to one estimate by the UN Food and Agriculture Organization. More conservative estimates put it at up to $20 billion. The difference exists because what qualifies as subsidy is a matter of dispute. Industrialized countries alone gave $6.7 billion of subsidies to their fishermen in 1996, according to the Organization for Economic Cooperation and Development (OECD). This fell to $6.2 billion in 1999, as more countries realized that subsidies might be the root of the overfishing problem, but rose again to $6.7 billion in 2003 (the most recent year for which data is currently available).

Top of the subsidies league is Japan, which handed out $2.3 billion to fishermen in one way or another in 2003. It is followed by the European Union, which hands out $1.74 billion, excluding Belgium and the Netherlands because they were slow with their figures that year. In third

continued

place is the United States, which doles out $1.29 billion. Individual EU members—Spain, France, Ireland, and Italy—all make impressive showings in their own right, doling out $503 million, $178 million, $68 million, and $180 million, respectively. Norway, another big player despite its tiny population, hands out $143 million, which works out to much more per head. Korea provided $495 million in 2003.

Subsidies amount to a staggering proportion of the actual value of landings. In the United States it was over 37 percent in 2003, in Japan 24 percent, and in the EU 22 percent. Two decades ago the Soviet Union handed out free boats and fuel to its fishermen—a big but unquantified subsidy. There remains a total lack of transparency everywhere about fishing subsidies and who gets them. The only public audit of them ever conducted, by the European Court of Auditors, found numerous payments to people who did not qualify for subsidies, including at least one to the owners of a vessel that had sunk.

What qualified as subsidies in the OECD's study included direct payments, such as grants for building new vessels and for making old ones safe, price supports for fish, grants for companies to set up joint ventures with other countries, tax exemptions and interest rebates for the purchase of vessels, and general services, such as state expenditure on fisheries science, research, management, and law enforcement. Buying access for fishermen to other countries' waters, from the north of Norway via Africa to the Falkland Islands, accounted for a sizeable

$193 million of the European Union's $1.7 billion expenditure on subsidies. Some countries now insist on recovering a significant percentage of the government's costs from the fishing industry. New Zealand, Iceland, and Australia, for example, recover 51 percent, 44 percent, and 39 percent, respectively, of the cost to the public of fisheries research, management, and enforcement. These countries, arguably, run some of the best-managed fisheries, so there seems to be a connection between no subsidy and good management.

The converse is also true. Subsidies create overcapacity in the industry. The global fishing fleet is estimated to be two and a half times greater than needed to catch what the ocean can sustainably produce. A big fleet creates pressure on politicians to use subsidies to buy licenses for fishermen to fish elsewhere if their own waters are overfished. In Europe it is those countries that subsidize their fishermen the most that lobby for this overcapacity to be exported to the rest of the world. (Spain, for example, spends $126 million on decommissioning vessels, which allows the modernization of its fleet.) Spanish and Portuguese fishermen, who used to be subsidized to fish in Moroccan waters, were *compensated* by the European taxpayer to the tune of another $244 million when they were kicked out after Morocco refused to renew one of the EU's controversial fisheries agreements.

Some people believe that there are good and bad subsidies. Half of all fishing subsidies received by Spain from the EU are environmentally damaging, according to one

continued

WWF report. The authors of this report strained their credibility by finding that 36 percent of subsidies were actually good for the environment and 15 percent were neutral. Generally, I believe that if you look hard enough, you find that a subsidy nearly always harms the environment or disadvantages someone somewhere, usually by distorting prices in favor of fishermen in the North, which penalizes the South. Even so-called neutral subsidies, such as those paid to improve the safety of boats, mean that fishermen will stay out longer in bad weather and catch more fish.

So what lies at the root of a democratic politician's impulse to dish out subsidies? First is a disgraceful need to buy votes with other people's money, often dressed up as the redistribution of wealth. Second is the misguided belief that subsidizing fishing is somehow *investing* in the industry. In fact, in a hunter-gatherer economy, you invest only by leaving the resource alone. The way to tackle subsidies in well-governed countries is to create transparency, a free press, and proper scrutiny by public auditors. That way people get to realize that farmers and fishermen are walking off with their money for no good reason.

One of the commendable aspirations of the round of free trade talks that began in Doha in 2001 is to reduce fishing subsidies of all kinds. Despite the addiction of most environmental bodies to obstructing the World Trade Organization's processes and carping about its failures, most recognize that this would be an unreservedly good thing.

9

LAW AND THE COMMONS

Atlantic Dawn is an evocative name for a vessel. To most Irishmen there is something odd about it, too. If you live on the mainland of Ireland, the sun comes up over the land, and east is never over the Atlantic, unless you are at sea or on an island. Kevin McHugh, the owner of the world's largest supertrawler, began his career as a fisherman-entrepreneur on Achill Island, off the west coast of Ireland, so quite conceivably the dawn did come up over the Atlantic for him. I remember reading in the *Irish Times* that when his beautiful new boat was built, McHugh planned to take the first opportunity he could to anchor off Achill Island so that his mother, Nora, could take pride in his new venture. There was great charm in the portrait of this unassuming, private, soft-spoken man who had become Ireland's most successful fisherman. You did have to remind yourself that the object of his pride was the greatest fish-killing machine the world has ever seen, a vessel of over 15,000 tons with a crew of a hundred to catch, pack, and freeze the fish that were stored in its cavernous hold.

Kevin McHugh made a career by spotting opportunities. He bought his first boat, the 65-foot *Wavecrest*, in 1968, at the age of twenty-one, and eight years later was the proud owner of the *Albacore*, built partly to his design, at a cost of $2.1 million. The closure of the Irish herring fishery led him to Killybegs, on the Donegal coast, where he concentrated on another pelagic fish, the mackerel. Then, like many other Irish skippers, he moved to bigger and bigger boats until in the late 1980s he placed an order many thought would break him, for a $21.4 million vessel that could steam even further to meet the demand for fish. The *Veronica*, named after his wife, was a highly efficient purse seiner designed to catch fish inside and outside European waters as fisheries near to home began to drift into crisis. On the way, however, he suffered a setback. The original *Veronica* was destroyed in a fire in Harland and Wolff's shipyard in Belfast while being repaired. It was eventually replaced, in 1995, with a larger *Veronica*, 345 feet long and, at 5,726 tons, some 1,280 tons greater than its predecessor.

With the *Veronica* fishing off Mauritania, McHugh appears to have seen further opportunities in the plentiful stocks of sardinellas and other pelagic fish off West Africa, already pursued by a Dutch fleet of vast superseiners. The vessel he commissioned to compete with them, the *Atlantic Dawn*, was built in Norway, making use of the subsidies available there for shipbuilding. The giant 480-foot purse seiner and pelagic trawler was built with $7.2 million in subsidies from the Norwegian government against a total cost of $89.3 million. It was designed to catch, process, and freeze up to 440 tons of fish a day and to accommodate 7,700 tons in its hold. McHugh told the *Irish Times* he was unhappy with the perception that such powerful ships posed a threat to the world's fish stocks.

Where we will be working, off Mauritania, there are strict controls and very strict arrangements, whereby we must take a certain number of Mauritanian fisherman on board, and an observer. They like to see us coming, because we are paying for our investment and there is a direct return to the state. Those waters down there are alive with fish—like Bullsmouth—but the air and sea temperatures make it very difficult to land quality. That's what we aim to do, and to do so in cooperation with the Mauritanians, so both of us can benefit.

One of the reasons for the vessel's size, he explained, was about having the horsepower to drive an onboard factory with sufficient freezing capacity to ensure the quality of the catch. His company would be selling the fish on the African market, a far better thing than the Russian vessels that were catching the same fish for fish meal. McHugh's financial package was put together by a syndicate of Irish banks.

The only problem came when the *Atlantic Dawn*, all sleek and lovely, was about to sail triumphantly for Dublin and it was revealed that it did not have a license to fish in EU waters—or any other waters, come to that. As the Irish Green party member of the European Parliament, Patricia McKenna, put it at the time: "Mr. McHugh has essentially said to the Irish Government: 'I have this boat, please find me some place to go fishing.' " McHugh was not alone in making that request. The syndicate of banks that had lent him the money was asking, too.

The *Atlantic Dawn* was a big project for a small country. People in government circles say it had developed such momentum it could not be allowed to fail. The Irish government did its best to find it somewhere to fish. It applied at the end of 2000 to increase its pelagic fleet on the grounds that there were as yet underexploited fishing opportunities in West African waters. The European Commission pointed out that Ireland's domestic pelagic fleet was already 40 percent larger than it should have been under a Europe-wide fleet limitation to which member states were supposed to adhere. This, the commission said, was illegal under EU law. No increase could be given unless the Irish authorities acted to clear the excessive fishing power of the pelagic sector.

While all this back-and-forth was happening, Frank Fahey, minister of the marine, gave the *Atlantic Dawn* a place on the Irish merchant register—though EU rules say that fishing vessels are required to be registered on member states' fishing register—and granted it several temporary fishing licenses. So the largest fishing vessel in Ireland went fishing as a merchant vessel for a year and a half. This, as the European Commission pointed out to the Irish government who had sanctioned all of this, was illegal, and in November 2001 the commission opened infringement proceedings against Ireland. As an Irish television investigation subsequently revealed, even the *taoiseach* (prime minister), Bertie Ahern, intervened in an attempt to break the impasse. He wrote to the head of the European Commission, giving the impression that if the *Atlantic Dawn* company was to go bust, it would have serious consequences for peace on the troubled border with Northern Ireland. Representations were even made to the EU fisheries commissioner, Franz Fischler, by Ireland's member on the European Commission, David Byrne, responsible for consumer affairs, who admitted that he had raised the matter with Fischler but stated that it was in the normal course of business and he had done nothing wrong. The result of all this was that, to the dismay of a number of EU officials who pointed out that national fleets were supposed to be getting smaller rather than larger, a deal was made that got the *Atlantic Dawn* onto Ireland's register of fishing vessels. The EU once again demonstrated how it appeared to be more interested in the regional politics of its member states than the sustainability of its fish stocks.

Ireland, however, did rectify what the commission called the "overhang" in its fleet capacity by removing the *Veronica* from its fleet and de-rating the power of another two fishing vessels. Though the *Veronica*, prior to the *Atlantic Dawn* the largest fishing vessel in the Irish fleet, was not actually going to stop fishing, the *Atlantic Dawn* was allowed onto the EU register. The reason given by a senior commission official for allowing Ireland more scope to fish was that scientific advice showed there was plenty of pelagic fish to catch in Mauritanian waters—the Dutch superseiners hadn't been taking up their quota. The *Veronica*, totally legally, now fishes in Mauritanian waters under a Panamanian flag, on a license owned by a Mauritanian company.

Dr. Euan Dunn, a fisheries expert working for Britain's Royal Society for the Protection of Birds, believes that the deal that allowed the *Veronica* to carry on fishing under a flag of convenience was a disgrace. Under a UN Food and Agriculture Organization action plan and under its own community action plan, the European Union was already committed to discouraging EU vessels from registering under flags of convenience that fail to fulfill their flag state responsibilities. Panama used to be one of the worst offenders against international fisheries agreements among the flag-of-convenience states, and although it has taken steps to remove some illegal tuna boats from its register, it must be seen as on probation at best.

This is not the only time that the European Commission has allowed the real killing power of the European fleet to expand. In another example, the commission accepted the size of the

Dutch pelagic fleet at twice the size it was supposed to be under a previous EU decommissioning round—unleashing a total of 103,400 tons of technologically advanced fishing capacity, representing several highly equipped supertrawlers and purse seiners, upon the waters of the North Atlantic and the rest of the world. It might, understandably, have been trying to keep control over a fleet fishing mostly in Mauritanian waters, which would otherwise simply defect to flags of convenience. But as far as reducing and containing fishing capacity goes, as Michael Graham and E.S. Russell first tried to persuade Europe to do sixty years ago, it has achieved very little.

What conclusions are to be drawn from the intriguing episode of the *Atlantic Dawn* and the *Veronica?* First, fishing registers in developed countries of the world, now all supposed to be limited and contained under an FAO action plan, in fact leak like a sieve. (Old habits die hard. Not long ago, there were actually subsidies for moving off the EU register to a flag of convenience.) Second, fishing capacity continues to grow everywhere. The Atlantic Dawn Company, which owns the vessel of that name and the *Veronica,* now advertises anchovies caught in European waters by the *Atlantic Dawn,* as well as a variety of pelagic fish caught off West Africa. If proof were needed that the *Atlantic Dawn* is contributing entirely legally to the global destruction of fish stocks, it is to be found on the company's Web site. One of the species on offer is the spectacularly overfished blue whiting.

It was assumed until very recently by the FAO that fishing vessels switching to "open" registries, where you can buy registration in return for a nominal fee, were old vessels, fully depreciated and nearing the end of their productive lives. Recently, however, someone in the FAO woke up and actually asked

Lloyd's Maritime Information Services about this, and discovered it was not the case. An increasing number of young and newly constructed vessels have moved to "open" registries. Ironically, the European Commission, which pushed the *Veronica* onto a flag of convenience, has pointed out that fishing by flag-of-convenience vessels represents a threat to the survival of fisheries worldwide.

So what exactly is the problem with flags of convenience? Ship owners have used them for years to avoid taxes and to cut labor costs, safety requirements, and training for crews. Cutting corners on health and safety hasn't gone down too well, particularly after some messy tanker accidents, but Panama and Liberia now have vessel inspections as stringent as anyone else's. By and large, the world has gotten used to a low-cost, lowest-common-denominator market in shipping. Indeed, some European countries, such as Germany and the Netherlands, even operate second registries with tax-free status on the grounds that if you can't beat 'em, why not join 'em?

The problem with flags of convenience is that they often allow fishing vessels to avoid conservation agreements for fish. Under the UN Convention on the Law of the Sea, ships sailing the high seas are subject only to the jurisdiction of the flag state. Only within a country's 200-mile limit or in other exceptional circumstances may a vessel of another country board or otherwise inspect a ship on the high seas. The flag state must therefore enforce all conservation agreements, including fisheries conservation and management agreements by regional fisheries organizations. Some do, some don't. Arguably, there isn't a problem as long as the flag state observes the fishing agreement. At least, that is what's claimed by the Spanish tuna fleet, which sails entirely under flags of convenience. Julio Morón, assistant director of the Spanish purse seiners organization, told me that the Spanish fleet is under flags of convenience purely for the tax advantages. People who have fished on Spanish purse seiners in the Indian Ocean say otherwise. They say a flag of convenience enables skippers to follow the fish unencumbered by the bureaucracy and quota that an agreement with an EU country involves. A skipper can make an arrangement with a particular African country on the spur of the moment and negotiate his own private fishing arrangements by fax.

I heard of one skipper in charge of a Spanish-owned vessel who thought he had bought a license to fish in Somali waters. Somalia, you will recall, is the only country in the world without a government at present, so he did this by agreement with one of three warlords who have effective control. When the vessel began fishing for tuna the crew was horrified to find their vessel being overtaken by British mercenaries in a fast Russian speedboat. The mercenaries pulled alongside and told the tuna boat it would have to pay $248,000 into an account in Chelsea or it would be towed into Mogadishu and impounded, which meant effectively stripped. Money was transferred very fast, and the British motored away. Apparently the purse seiner had contacted the wrong warlord for the area it was fishing in.

There is always somebody prepared to accept money in return for allowing a vessel freedom to fish. A change of registration can be carried out by fax, over a satellite phone, while at sea. At the bottom of the cascade of money and responsibility are many flag states that either do not enforce the treaties they have ratified or never signed them. A basic principle of international law is that if a country does not adhere to a treaty, it is not bound by its provisions. A vessel flying the flag of Belize, for example, may fish perfectly legally for tuna in the Atlantic without paying any attention to conservation measures laid down by the International Commission for the Conservation of Atlantic Tunas. This amounts to a gaping loophole in the law, which appeals to those who want to fish on the high seas with impunity, and to those who wish to fish illegally in other countries' exclusive economic zones under the appearance of fishing on the high seas. The popularity of particular flags changes as international pressure is brought to bear on certain countries to remove miscreants from their registers—as ICCAT has made certain Central American states do through trade sanctions. But there remains a handful of flags that appeal to companies that gear up to fish illegally for high-value species because they calculate that the risks of being caught are outweighed by the profits if they get away with it.

Pirate fishing of this kind can be every bit as cruel as the pirating of popular lore. Hélène Bours of Greenpeace International was sailing off the coast of Sierra Leone in the MV *Greenpeace* on September 11, 2001. She had already seen several vessels in a very dilapidated state, apparently lacking the most basic safety equipment and with the crew living in deplorable conditions. That morning the MV *Greenpeace* picked up a mayday call from a ship in distress: the *Estemar 5*, a Korean-owned vessel registered to the Tikonko Fishing Company of Sierra Leone. The Greenpeace vessel headed immediately to its aid. Bours recalled:

Shortly before we reached the location given in the distress call, about 2½ hours later, we saw a large slick of oil in the water, as well as some mooring ropes, plastic boxes and many other types of debris—presumably from the ship in distress. We set up look-out groups on the bridge and in the crow's nest and steamed about looking for survivors. A few hours later, a small boat with an officer of Sierra Leone's navy informed us that nine crew members of the fishing boat were missing. Only the captain and one deckhand had so far been found alive. We launched two of our inflatable boats to expand the search and to have a better chance of seeing any survivors. After another hour, without having seen any survivors or bodies, we terminated the search operation. Astoundingly, most of the other fishing vessels in the area had not even stopped fishing to help.

The *Estemar 5* was a serial offender. It had been sighted five times in a year in prohibited areas of the Sierra Leone 200-mile limit during flights carried out by the Surveillance Operations Coordination Unit in Gambia.

Very slowly the gray areas of the law of the sea are turning to black and white. The 1995 UN fish stocks agreement, which only came into force in 2003, requires flag states actively to comply with fishing agreements on the high seas, rather than merely not undermine them. It also confers considerable powers on regional fisheries organizations to inspect and carry out surveillance of vessels fishing in their area, which includes allowing them to insist that satellite transponders be fitted. A voluntary plan of action agreed by the international community in 2001 provides support for flag states by tightening up scrutiny of the vessels on their registers and of port states—countries where the vessels dock—that collect information on fishing and deny access to flag-of-convenience vessels. As yet, however, the law of the sea still lacks the power to allow anyone but the flag state to impose measures on vessels on the high seas. Nor is there a way of forcing countries with open registers to observe the requirement, repeatedly stated in international law, that there should be a "genuine link" between fishing company and flag state. That would mean that offenders could be easily prosecuted.

No edges are sharp when it comes to the law of the sea because it is always in someone's interest, even in the most apparently well-governed nations, for them to be fuzzy. The consensus treaties that make up the law of the sea are full of loopholes because they were negotiated that way. It is an ancient quandary, so a little history might make things simpler to understand.

The Romans believed that what no man controls, no man can own. Justinian, writing in the sixth century AD, said that the air, flowing water, the sea, and the seashore were common to all. In the fifteenth and sixteenth centuries Europe's crowned heads made elaborate and unenforceable claims over vast areas of sea and land. Venice claimed the Adriatic; England the North Sea, the Channel, and part of the Atlantic; Spain the whole Pacific; Portugal the Indian Ocean and much of the Atlantic. Papal bulls, which Spain and Portugal claimed as their authority for laying claim to huge tracts of ocean, were unpopular and attracted opposition from nations who asserted the right of free passage. The case against these papal bulls was stated by Hugo Grotius, a Dutch jurist, in a 1609 essay entitled *Mare Liberum* (The free sea). He insisted, on the basis of classical legal precedent, that the sea was *res communes* (common to all nations). John Selden, an English lawyer, historian, and antiquarian, riposted on behalf of the English crown with a 1635 critique called *Mare Clausum* (The closed sea), which asserted the state's right to claim sovereignty over the seas adjacent to its territory. The Dutch herring fleet's practice of fishing within sight of the English shores is understood to have been one of his reasons for doing so. In fact, by 1625 Grotius was already conceding that the common-sea principle might not apply to the adjacent sea. The distance over which the state was held to exert effective control became established in the eighteenth century as the distance a shore battery could fire, then 3 miles. As gunnery improved, there was no attempt to extend this range, although some countries asserted sovereignty over 12 miles to police matters such as smuggling. The 3-mile limit lasted into the twentieth century.

The tension between the common sea and the closed sea remains, however, long after the 1982 UN Convention on the Law of the Sea laid out the legal basis for each country to establish a 200-mile EEZ. (If the distance between nations is less than this, the EEZ stretches to the median line between the two, e.g., down the middle of the North Sea.) The 1995 UN Straddling Stocks Agreement, which allows members of regional fisheries organizations to inspect vessels, has extended enforcement of a kind to the high seas. It is now theoretically possible, through satellite tracking and overflights by fisheries inspectors, to see and apprehend vessels that are fishing in the wrong place—much more like the system of de facto power that coastal states exercise in their own waters identified by Selden—though not to enforce them on nations that are not party to the agreement. But although some fisheries, for squid, tuna, and toothfish, are now global, regional fisheries bodies remain exactly that—regional—and lack power. In large areas of the ocean beyond the 200-mile limit, such as on the Madagascan Ridge, there are rich fisheries with no regional organizations at all, and anything goes.

What happens in areas of ocean such as this is the purest form of what has become known as the "tragedy of the commons," after an essay published in *Science* in 1968 by Garrett Hardin, an ecologist at the University of California, Santa Barbara. Hardin claimed modestly that he had merely renamed a phenomenon first described by the nineteenth-century British political economist William Forster Lloyd, who mused over the devastation of some common pastures in England compared with enclosed land.

The tragedy of the commons, as defined by Hardin, works like this. A pasture is open to all. Assuming each human exploiter of the common is guided by self-interest, he will try to keep as many cattle as possible on the common. This arrangement may well work for years because war, rustling, or disease will keep the numbers of men and beasts below the carrying capacity of the land. At the moment when man and environment come into equilibrium, tragedy begins. Each herdsmen will ask himself what he is to gain by adding one cow to the common, compared with the loss he will make by overloading the pasture. The gain will always be greater to him, and the loss, because it is shared, less. And so it goes on, with every herdsman buying himself another cow until the common property is ruined and the cows starve.

Those who nod sagely and quote the tragedy of the commons in relation to environmental problems from pollution of the atmosphere to poaching of national parks tend to forget that Garrett Hardin revised his conclusions many times over thirty years. He recognized, most importantly, that anarchy did not prevail on the common pastures of medieval England in the way he had described. The commoners—usually a limited number of people with defined rights in law—organized themselves to ensure it did not. The pastures were protected from ruin by the tradition of "stinting," which limited each herdsman to a fixed number of animals. "A managed commons, though it may have other defects, is not automatically subject to the tragic fate of the unmanaged commons," wrote Hardin, though he was still clearly unhappy with commoning arrangements. As with all forms of socialism, of which he regarded commoning as an early kind, Hardin said the flaw in the system lay in the quality of the management. The problem was always how to prevent the managers from furthering their own interests. *Quis custodiet ipsos custodes?* Who guards the guardians?

Hardin observed, crucially, that a successful managed common depended on limiting the numbers of commoners, limiting access, and having penalties that deterred. The important thing about a successful common in medieval England is that there was not free access. The number of commoners on an English common were often as few as a dozen. Hardin points out that even when herdsmen understood the consequences of their actions, they were generally powerless to prevent damage without there being an adequately coercive way of controlling the actions of each individual.

If any group should be able to make commons work on the basis of Marx's principle of "from each according to his abilities, to each according to his needs," Hardin said, it should be an earnest religious community, such as the Hutterites in the western United States. Hardin observed that when the number of people in the community approached 150, individual Hutterites began to undercontribute according to their abilities and overdemand for their needs. He concluded that "numbers were the nemesis."

At sea, where a common exists in most waters governed by nations or regional fisheries organizations, stinting remains the most favored form of management. Where stinting doesn't work, which is in more places than not, this is generally because there are too many people involved for there to be trust between fishermen, or the penalties are not sufficient to deter them from following their financial interests. If there is to be a way of successfully managing a common sea, there needs to be a strong chance of being caught and tough penalties when an offender is convicted.

None of Hardin's requirements for a successfully managed common is fulfilled by high-seas fishery regimes. Nor very often do they work within 200-mile limits, either. The declaration of such limits still gives each nation the problem of allocating rights and setting adequate penalties. Frequently the share is resented if it is thought that some individuals have begun to cheat. When the well-meaning commoners, the bedrock of a working community agreement, perceive that others are cheating, they begin to cheat, too. One of Hardin's lessons is that a common can evolve from a managed one to a tragic one. "With an unmanaged common," he wrote, "ruin is inevitable." A ruin with which the world is particularly familiar even contains the word *common* in its title—the European Union's Common Fisheries Policy.

In later years Hardin confronted the fisheries problem directly. He wrote: "If each government allowed ownership of fish within a given area so that an owner could sue those who encroach upon his fish, owners would have an incentive to refrain from overfishing." Hardin was an early and influential advocate of granting property rights to fishermen to persuade them to conserve the resource and to police each other.

There can be little doubt that one of the most perfect examples of an unmanaged common on the planet is the southern Oceans around Antarctica. There ruin has beckoned for two species of toothfish, the Patagonian toothfish and the Antarctic toothfish, both large, bottom-living predatory species that can grow up to 6 feet long and live for fifty years. Antarctic toothfish are a less numerous species limited to the Ross Sea and shelves around Antarctica. The reason stocks have been in trouble is not that there are no rules but that the rules are so difficult to enforce in a vast, remote ocean with the highest winds and largest waves on Earth.

As with other relatively inaccessible deepwater stocks, fishermen turned to the toothfish only after other valuable Southern Hemisphere fish, such as the austral hake and golden kingclip, were eclipsed. Once consumer resistance was overcome by renaming it Chilean sea bass—perfect washed down with a glass of Chilean chardonnay—Patagonian toothfish sold well in the white-tablecloth restaurants of the United States and Japan. Its flaky white flesh had a pleasant flavor and stood up to being overcooked as robustly as cod did. The price of toothfish has risen on the back of limited supply, since fishing for toothfish peaked in the mid-1990s, wholly justifying its description as "white gold."

The rewards for catching toothfish are very high. The risks of being caught are very low. On the back of the current price, it is possible for a toothfish poacher to catch enough to pay for his boat, pay his crew, and make $500,000 profit in a single trip, according to Dr. David Agnew of Imperial College, London, who manages the toothfish stocks around South Georgia, a British dependency in the South Atlantic Ocean, on behalf of the island's government. No wonder illegal operators are willing to risk their vessels being impounded. If that happens, they do not hang around to get them back.

With a peak of at least thirty-three large vessels fishing illegally in the southern Indian Ocean each year, and only one or two arrests a year, "the odds were better than people smuggling or drug running," according to David Carter, chief executive officer of Austral Fisheries, based in Perth, Australia. Carter's company runs two boats fishing legally for toothfish. In Australian waters poachers met a new phenomenon: legal fishermen with ownership rights to their fish who were prepared to campaign to protect their assets.

With the toothfish two major ecological problems have intertwined, which have helped the legal fishermen to make a case for tougher action by governments. First, some stocks of toothfish rapidly became chronically overfished, with illegal fishing taking more than half the annual catches (80 percent by some estimates). Second, the long-lining methods used by the fishermen who fish for it illegally or by using flags of convenience in the southern oceans have been wiping out albatrosses; in fact, seventeen of twenty-four species are now under threat of extinction. Australia has dealt with bird bycatch in its waters— 200 miles around its remote southern territory of Heard Island—by banning long lining in favor of trawling until it could be demonstrated that long lining can happen with significantly lowered incidental catch of seabirds. Positively, trials of bird mitigation measures on long lining in Australian waters over the past four years have reduced bird kills from that method of fishing to nearly zero, demonstrating what can happen when industry has an incentive to work with scientists, managers, and government. Similarly, the island of South Georgia has cut bird deaths to fewer than twenty a year by requiring long lines to be released at night or under multicolored deterrent streamers. In the high seas it is up to the fishermen themselves what measures to take, and, being unregulated or illegal, many choose to take none.

The plundering of the seas around Antarctica is not new.

British and Norwegian whalers killed 30,000 blue whales in their most successful year, 1929–30. Now these gentle giants, the largest creatures on Earth, are reduced to a total global population of 1,500 at most. The Soviet fishing fleet, once the largest in the world, used to fish in the southern oceans for a variety of stocks, including icefish and the marbled rock cod, the latter a valuable, beautiful, and slow-reproducing fish that used to exist around the British dominion and old whaling base of South Georgia. By the early 1980s, the entire stock had collapsed.

Until the British declared a 200-mile limit around South Georgia and the South Sandwich Islands in 1994 and began to enforce it with a new armed fisheries protection vessel, the *Dorada*, there was a toothfish free-for-all around South Georgia. After the British got a grip on that, using satellite surveillance and overflights from the Falklands to lead the *Dorada* to the poachers, the problem went elsewhere—to the poorly enforced 200-mile zones around the sub-Antarctic and Antarctic islands under the jurisdiction of South Africa, France, and Australia.

Even so, the first time many people heard of the hunt for toothfish poachers was when the story came out, in August 2003, of the marathon 4,000-mile pursuit of the Uruguayan-flagged toothfish vessel *Viarsa* by vessels of three nationalities. The *Viarsa* was spotted fishing near Heard Island and took flight. Believed to be carrying an illegal haul of toothfish worth $3.8 million, the vessel ignored repeated radio orders to stop and pressed on west, through mountainous seas, toward its home port of Montevideo, pursued by a lone, unarmed Australian customs ship, the *Southern Supporter*. For fourteen days the customs ship plowed on in pursuit of the toothfish pirates, through a blizzard and packs of icebergs, often losing visual bearings on the faster fishing vessel in huge seas and foul weather. It was later joined by a helicopter-equipped South African icebreaker, *Agulhas*, and by the South African salvage tug *John Ross*, manned by armed fisheries protection officers and capable of 20 knots. It was not, however, until the ship was intercepted by the Falklands-based British vessel *Dorada*, equipped with a machine gun and two fast inflatables, that the fleeing vessel was boarded and its forty crew members arrested.

What the pursuit of the *Viarsa* by an unarmed vessel showed is that Australian enforcement of its Antarctic fisheries was until then "almost laughable," says David Carter of Austral Fisheries. The same, he believes, was true of the French waters around Kerguelen Island and Crozet Island, and South African waters around Prince Edward Island and Marion Island, though there has recently been a spate of public criticism about seabird deaths in French waters. Carter's company owns the right to catch 71 percent of the annual legal quota of toothfish in Australian waters in perpetuity. He believes this arrangement has been instrumental in his company's vigorous defense of its rights. He and his fellow legal fishermen created the Coalition of Legal Toothfish Operators (COLTO). In eighteen months they set up a Web site and published a rogues' gallery of the illegal vessels their boats saw and photographed on the fishing grounds. He believes this would not have happened if they did not have property rights to defend. "As a result of that interest, we have an enlightened self-interest in making sure those resources are protected. We have spent a lot of time, energy and passion in making this debate what it's been."

COLTO's members were distinctly unimpressed by the ability of the Hobart-based Commission for the Conservation of Antarctic Marine Living Resources (CCAMLR, pronounced "kammelar") to take effective action. CCAMLR, set up by

Antarctic Treaty nations in 1982 largely as a scientific management body, has many virtues on paper and in its treaty, which is one of the most enlightened where fisheries are concerned. (It is one of the few, for example, that explicitly recognizes the need for ecosystem management, acknowledging that if Antarctic krill, which now have few ready markets, become overfished, the whole terrestrial ecosystem of seals and penguins could be faced with collapse.) For all CCAMLR's virtues, one of them is not enforcement. Being a consensus organization on the UN model, like the equally pious and ineffective FAO, it initially declined COLTO's exhortation to blacklist persistent offenders. So COLTO took matters into its own hands and compiled the blacklist itself. Over the past three years, CCAMLR has introduced its "IUU list" of offenders and, while not perfect, is starting to have a positive impact at reducing illegal, unregulated, and unreported fishing yet further.

The rogue's gallery identifies at least two groups that have taken illegal fishing beyond the realm of shameless opportunism. *Viarsa* and its sister ship *Arvisa* were two of twenty-six vessels owned and operated by what COLTO called the "Galician Syndicate," based in northern Spain. Many of the Galician Syndicate's fishing vessels were registered in Uruguay, others in Ghana, Argentina, Belize, and Panama. Until the arrest of the *Viarsa*'s crew, there seems to have been great tolerance shown by Uruguayan officials to known poachers remaining on their registers, even generating valid CCAMLR catch documentation for them and accepting bogus satellite positioning data. Since the arrest, the *Viarsa* has been deregistered in Uruguay, and many changes have occurred within the Uruguayan management and control system to better enforce international regulations and requirements. The *Viarsa*'s officers were found not guilty of illegal fishing by a jury in Australia in late 2005—the main legal issue being the inability of the prosecution to prove that the boat had a line in the water and was fishing for toothfish at the time the patrol boat came across her. A civil case (against the boat, gear, and catch) proceeds, but the whole episode illustrates the difficulty of applying sanctions in some of the remotest places on Earth.

While flags of convenience are the main problem in trying to regulate the toothfish trade, ports of convenience are a problem, too. Ports where toothfish vessels take on supplies included Port Louis in Mauritius, Durban in South Africa, Walvis Bay in Namibia, and of course Montevideo in Uruguay. Since Mauritius, South Africa, and Namibia have all taken steps to tighten up controls on illegal and unregulated landings, the poachers have taken to transshipping at sea. Once the toothfish has been transferred to a mother ship, as part of a huge cargo on its way to multiple ports in Asia, port state control becomes ineffective. This has been a major loophole in the Catch Documentation Scheme introduced by CCAMLR (which became binding in 2000), as has the dilatory behavior of Canada and the EU, among others, in putting into force at home what they agreed to overseas. Moves to address this have been increasingly successful, with the United States requiring electronic catch documents to be completed and paperwork to travel with every shipment of toothfish, no matter what port it travels through. The authorities have worked hard to keep a step ahead of the illegals, constantly changing focus to plug the loopholes. But all this effort makes you wonder whether it would not be better to change international law so that only those who are licensed by the regional fisheries management organizations (in this case CCAMLR) are allowed to catch and sell fish taken on the high seas.

The result of the *Viarsa* arrest has been a political break-through for COLTO in Australia. Illegal fishing now has what David Carter calls "taxi-driver recognition" as a major problem. Because its customs vessel was unarmed and could not make the vessel stop, the arrest of the *Viarsa* cost the Australian government $3.7 million. Similar arrests are now seen as a false economy. Australia has, however, been getting serious about enforcement and copying the British example in South Georgia. The government initially promised that $75 million will be spent over two years in patrolling the area with a new vessel armed with a 50-caliber machine gun and capable of mounting armed boarding parties. Within twelve months this was recognized as insufficient to complete the job, and the Australian government made a further $158 million available over a three-year period for more patrols, as well as to share the costs of satellite surveillance systems with France and South Africa. France has also been getting serious about policing around the Kerguelen and Crozet Islands, where toothfish stocks are estimated to be only a quarter of what they once were. It has signed a treaty of cooperation with Australia, implemented satellite surveillance of its waters, and converted two captured illegal toothfish boats into civilian patrol vessels for its waters (with operating funds provided by the legal French industry members), along with use of one naval vessel permanently on station in the sub-Antarctic. South Africa is working with France and Australia and has built and released three new patrol vessels, one of which is designed to withstand sub-Antarctic conditions, for chasing illegal toothfish poachers. In less than five years, these remote sub-Antarctic territories has become regularly overflown by patrols, constantly scanned by satellites, and monitored by fisheries protection vessels. Prosecutions and apprehensions are more frequent. The waters within 200-mile limits are being patrolled more effectively.

Meanwhile, however, the problem has just got bigger for the toothfish. Tagging studies have shown that eleven tagged toothfish did not stay conveniently within Australian jurisdiction but swam some 1,000 miles to Crozet Island, in French territory, across seas that are open for anyone to fish if they are using a flag of convenience. This means the population outside the 200-mile limits over which Australia and France have jurisdiction is likely to be significant and in need of management. This in turn has highlighted the inadequacies of CCAMLR and other international management bodies, which are hamstrung by the concept of the freedom of the high seas. CCAMLR was set up as an environmental body, not a fisheries surveillance and conservation enforcement organization, and is all that stands in the way of the ongoing destruction of a significant toothfish fishery in the Ross Sea, off Antarctica. It is not enough. In the words of David Carter: "That's the powerlessness of CCAMLR, and any other international regional management body. Ultimately, somebody's got to own those fish, otherwise it's the failure of the commons all over again. Flags of convenience are inevitably going to be used by unscrupulous operators to avoid regulations and control. We just have to find ways of making CCAMLR members pursue their own nationals who are fishing unscrupulously under flags of convenience. The concept of the freedom of the seas is a crock and needs to be changed before it's too late for high seas resources."

You need an ice-class vessel to fish in the Ross Sea, which means those unregulated vessels now fishing there in the Antarctic summer do so at some risk, as they don't care about the safety and protection of their crews any more than they do

about catch limits or seabirds. How long will it be before someone constructs an icebreaking fishing vessel, perhaps somewhere such as Norway or Japan, where the government still
dishes out shipbuilding subsidies, to take a firmer grasp of this
beckoning opportunity?

10

THE SLIME TRAIL

Mercamadrid, 6:00 a.m. Madrid's modern fish market is the opposite of Tokyo's historic Tsukiji and, as a result, the envy of many visiting Japanese. It occupies 82 acres of purpose-built covered halls and car parking off Madrid's N40 orbital motorway, and is serviced by roads, not by the sea. The first impression as you arrive is of a vast swarm of white vans belonging to the restaurants and shopkeepers of the city. Mercamadrid handles only 220,000 tons of fish a year, compared with Tsukiji's 660,000 tons, but that still makes it the largest fish market in Europe, the second largest in the world.

Although the Spanish capital is hundreds of miles from the coast, it is a fish magnet, and has been for centuries because of the Catholic monarchy. When the king was in the city, he and his court would observe the Church's formal requirement to eat fish at least two days a week (curiously, just what health experts now recommend). Madrid has long had an umbilical connection to the Galician ports. In the last century, when roads weren't as good, trucks racing to bring fish overnight to Madrid were often involved in spectacular crashes.

Mercamadrid is not particularly interested in introducing

new tastes; it is in the business of satisfying the capital's traditional demands from whichever ocean has the right fish. You very quickly notice that there is an awful lot of hake on the market—surprising, given that scientists tell us that hake is on the verge of collapse in European waters. Perhaps it's less surprising when you realize that hake is a mainstay of traditional Spanish cooking. So the market is full of hake—gunmetal-gray fish up to 3 feet long, and one or two boxes of suspiciously small ones, at or below the minimum landing size, which are a delicacy. When you ask where all the hake is from, the answer is the same: the north of Spain. This is the answer you get even if the hake is labeled as from Namibia, Argentina, or West Africa and not actually from Europe at all. This answer does not betray ignorance but caution. One detects in the stallholders a reticence to look beyond the port where the fish is landed for fear of discovering a secret that might be better left unknown.

Occasionally a case comes to court that brings to light the startling extent of illegality in the hake fishery and the tolerance of rule breaking in Spanish fisheries generally, evidence of regional cultures with a rapacious attitude to the sea and a striking disrespect for the conservation of stocks. There is northern hake, which is caught off Ireland and Scotland, and southern hake, which is caught in the Bay of Biscay. Southern hake is, if anything, in worse shape. But there's not much to choose between the two. There is very little European hake left. You do not need to look far to see why: catches that are well over quota, and completely inadequate enforcement of those quotas in Spanish ports. In 2003, in a case that confirmed many of the suspicions British fishermen entertain about Spanish fishermen, an Anglo-Spanish fishing company was fined almost $2 million for a fraud in which its skippers caught more than twenty-five times the amount of hake they declared over two years, making

$1.8 million in extra profits. The *Whitesands* was one of a fleet of Spanish-owned, British-registered vessels, based in Milford Haven in South Wales, fishing off Ireland and to the west of Scotland and landing into northern Spanish ports, such as Coruña and Vigo. Swansea Crown Court heard that successive skippers of the *Whitesands* fiddled logbooks with "breathtaking arrogance," declaring only 4 percent of their true catches. Alterations downplayed the amount of hake the vessel had caught over forty-six fishing trips by approximately 558 tons.

The Spanish company even carried on after the fiddle was uncovered and legal proceedings had begun. It then went into voluntary liquidation in a deliberate attempt to escape being prosecuted. Plymouth Shipping, which operated the trawler, and Santa Fe Shipping, which owned it, were convicted of twenty-seven counts of fisheries fraud, including the alteration of logbooks, failure to record species under quota, and submission of incorrect landing declarations. To date no one has paid the fines. That same week a fine of $893,000 was imposed on the Spanish-owned, Milford Haven–based *Grampian Avenger* and *Grampian Avenger 2* for underdeclaring 181 tons of hake.

The way the EU works is that the European Commission strives, not always that hard, to tighten enforcement against the sheet anchor of its member states, notably in the south of Europe. In a spirit of even-handedness, the commission recently issued proceedings against Spain and the United Kingdom for failing to enforce rules on fisheries. Commission inspectors found that Spain did not spend enough on its fisheries inspectorate to ensure the adequate inspection of fishing activities at sea or onshore. Lack of staff and lack of equipment were compounded by a confusion of roles between local and national authorities, incompatible electronic equipment and delays.

The commission noted failures to follow up observed cases

of false declarations. It found that in the Canary Islands, where the vessels fishing in African waters land their catches, inspections were directed primarily at foreign vessels, while Spanish vessels were exempted. Shortcomings were observed in the inspection of vessels fishing under the agreement between the EU and Mauritania. On the mainland inspectors witnessed landings, sales, and transport of fish without any control by the local authorities. When they checked data recorded by local inspectors over that period, they found that only 15 percent of the 220 tons of hake put on the market had been declared.

At La Toja, a classic Galician restaurant off the Plaza Mayor in Madrid, a waiter lays out hake brought directly from Vigo in the refrigerated window. On the lunch menu are also cod, grouper, lobster, red bream, salmon, sea bass, sole, squid, and turbot. House specialties include the classic *merluza à la gallega*, hake boiled with potatoes, garlic, and peppers. They also serve hake neck—the head of a large hake cut in half and grilled, which non-Spaniards may not find to their taste.

Galician recipes are simple, not sophisticated, explains Manolo Sertage, the headwaiter. They depend on the quality of their ingredients, so Madrid's top restaurants will pay that little bit more for the best. One has a sense that the clientele of La Toja will go on paying what it takes to eat what they have always eaten. At the heart of Spain, at the top of the food chain, everything rather splendidly remains the same. But a slime trail leads all the way from the sea to the people feasting in the city, something the feasters prefer not to think about.

"Black fish"—the name given to illegal catches—is Europe's guilty secret, not just Spain's, but Spain's part in it is the largest of all. I asked Harry Koster, head of the European Commission's fisheries inspectorate in Brussels, how bad the illegality was. He said that 60 percent of hake, landed mostly in Spain, was unrecorded—in other words, illegal. How much cod in Britain was illegal? Fifty percent. All official figures from the International Council for the Exploration of the Sea (ICES), all pretty disgraceful. Many of Europe's fish stocks are in danger of irreversible decline—every other cod, every other hake that hits your plate is stolen from the general public and their grandchildren, the rightful owners of the sea, all because no one has the political courage to enforce the rules.

One somehow does not expect the degree of criminality by fishermen in Europe's waters to be the same as that perpetrated by pirates in the seas around Antarctica, but it is, and that is still not readily understood when fishermen come on television with their tales of woe every time scientists propose a lower quota. The only difference between those who ply distant waters and fishermen who work inshore areas is that toothfish poachers make more money and do not have to go to the trouble of buying quota, licenses, and legitimate gear and observing rules on days at sea, which fishermen do in Europe before they can get down to fiddling. Toothfish pirates just happen to be cutting pieces off a (for now) larger cake.

The reason Spain is implicated so often in breaches of the rules is, one suspects, because of political pressures that have arisen because of its oversized fleet. This fleet fishes in more waters than any other European country. Unfortunately, what this means is that very often its boats break the rules, where they exist, on a global scale. Spain is not alone in breaking the rules outside EU waters. The Portuguese fishing fleet, though not as large, gives Spain a run for its money when it comes to illegality, particularly in the North Atlantic. The most spectacular example of what these two nations are up to is their flagrant illegal fishing on the edge of the Grand Banks, just outside the Canadian 200-mile limit, some twelve years after fishing for cod and several other species was banned. A decade ago, Canada caused a major international incident when the fisheries minister, Brian Tobin, authorized the arrest of the crew of a vessel, the Spanish-flagged *Estai*, that was fishing illegally for Greenland halibut just outside Canadian waters. The EU accused him of acting illegally, and since then Canada has tried to avoid restarting the dispute. It still has a controversial statute on the books that bans boats flying flags of convenience from operating on the banks outside its 200-mile limit, which is why vessels that fish outside that limit do so under EU flags, when previously they used a variety of foreign flags.

One of the legacies of the *Estai* incident is that the Northwest Atlantic Fisheries Organization (NAFO) now has some of the toughest rules of any regional fisheries organization, intended to prevent the Canadians from taking unilateral action again. These include the requirement that 100 percent of vessels fishing in the area must carry independent observers at all times. The reports of observers on some of these vessels make chilling reading. They show that on seventy-two days of 2002, EU vessels deliberately targeted several species for which fishing was banned, primarily American plaice and cod. For some reason the details of the observers' reports are never published or made use of by the European Commission, except to check up on inspections by member states. Only a bowdlerized version, missing many of the most damaging details so as not to cause international offense, is published annually in Canada. The observer reports simply gather dust in a drawer in Brussels. So, using Europe's freedom of information legislation, I decided to ask the European Commission for reports on certain vessels. To my surprise, they were duly produced.

It is vital to remember that we are not talking about an alleged crime of routine cheating or poor bookkeeping here, but a crime of deliberately steaming across an entire ocean to fish for stocks that are internationally recognized as having collapsed and on which there is a moratorium in the faint hope that they might one day recover. In economic and ecological terms, this is about as serious a crime as you get. It cheats not just present generations but future ones, who might have hoped to benefit from the fish. The observer reports I obtained from the commission revealed that a territorial war is going on between fishermen and inspection vessels in the waters administered by the Northwest Atlantic Fisheries Organization. Perhaps the most thought-provoking thing the reports show are the failures of enforcement by the authorities in Portugal and Spain, and the apparent official tolerance of illegal fishing.

I read, for example, that the Portuguese-registered stern trawler *Solsticio* entered NAFO waters on March 19, 2003. A British observer from the firm MacAlister Elliot & Partners was aboard. The vessel was watched carefully on its month-long trip by Canadian inspectors using the vessel's positioning data. The inspectors then boarded it on May 5. They reported finding in the tunnel freezer, concealed behind empty racks of fish trays, evidence of previous tows, mostly consisting of the moratorium species cod and plaice. While the inspectors were on board, the vessel moved into deep water but continued to trawl, so when the net was retrieved, it contained 1.3 tons of fish, 60 percent of which was Greenland halibut. This reduced the amount of cod and plaice to 40 percent, just legal as a by-catch. The observer's figures showed that out a total of 312 tons of fish for the trip, 65 percent was moratorium species: cod (91 tons), plaice (96 tons), and witch (deepwater sole, 15 tons). The vessel was reported to NAFO for "directing" on moratorium species, that is, fishing for them deliberately. From its satellite transponder records, it was identified as fishing for two weeks in shallow water where cod and plaice were found. It was reported catching moratorium species by an observer and found to have illegal species on board during an inspection. Yet when the vessel was inspected in port back in Portugal, the inspectors there found "no infringements" and said they discovered less than 10 percent illegal species aboard. How much faith can one have in such an enforcement system?

Two other Portuguese vessels, the *Calvão* and the *Lutador*, were reported for misreporting their catches and undertaking directed fishing of banned species in December 2002. Yet in both cases the Portuguese authorities chose not to recall the vessels to port for inspection. This lack of follow-up, say the Canadians, allowed both vessels to fish into a new quota year, thereby preventing identification of landed catch relevant to the 2002 inspection. The observer aboard the *Lutador* found that the vessel was catching and discarding small plaice while it was fishing for yellowtail flounder. The rules say that if more than 5 percent of a vessel's catch is a banned species, it has to move 5 nautical miles. The observer's statement indicates that after 5 percent of plaice were caught, the rest were simply discarded. An observer aboard the *Pescaberbes Dos,* a Spanish vessel, found quantities of undersized Greenland halibut caught on many occasions, and noted, "Most of these fish were retained, and on a few occasions the skipper did not move the regulatory 5 nautical miles." The observer, who is supposed to weigh the catches himself, reported that he was asked to fill in his weekly reports of catches according to the figures he was given. Four other Portuguese vessels were accused by Canadian inspectors of fishing for moratorium species. Some were even photographed using tarpaulins to conceal illegal catches from air surveillance patrols.

EU vessels, of course, are not alone in misbehaving in NAFO waters. Canadian air surveillance has observed Russian vessels on numerous occasions with liners in the trawl, that is, one net inside another, which gives a much smaller mesh size and catches greater numbers of juvenile fish. On August 14, 2003, Canadian inspectors approached the *Andrey Paskov*. As they approached, the vessel relowered the trawl under water. However, the cod end of the net floated to the surface and inspectors were able to see at close range a liner with an illegal mesh and a catch of 5.5 tons of redfish. The Russian vessel then released the entire trawl onto the ocean floor. The master of the *Andrey Paskov* was reported to NAFO for obstructing the mesh of his trawl.

It is difficult not to suspect someone at EU headquarters in Brussels—though not those who made the information available to me—of conspiring to keep the flagrant and disgraceful behavior of its vessels secret. Canadian officials find it frustrating to carry out inspections that are not acted upon by the flag state and to see observer reports that are simply filed by the EU. "When you return to port in Portugal and Spain with large quantities of illegal fish, nothing happens to you," said one Canadian official. "There are twenty nations fishing in these waters and only two do not land in Canada. Which do you think those are?"

I put it to a member of the European Commission that there was plenty of evidence from observers to prosecute the masters of several EU vessels, even where the port inspector seems to have missed the evidence, so why did this not happen? The official said the reason was that observers' evidence cannot be used on its own in court in Europe because it does not have any greater weight than the word of the master of the vessel. It does have more weight in Canada, where the observer is seen as a dispassionate figure whose evidence should, on balance, be believed. With European vessels, therefore, the burden of finding evidence rests with the port inspectors, who appear to be blind and deaf. Whether this anarchy will be improved by basing the new EU fisheries inspectorate in the Spanish port of Vigo remains to be seen.

Why is Spain so obviously the most ruthless fishing nation in Europe and possibly the world? Why is it also the most influential fishing nation in the EU? Partly because it has an oversized fishing fleet, as well as a distant-water fleet, when other countries, such as Britain, have given theirs up. But why does Spain have so many vessels? There lies a dark secret that more people should know the next time they tuck into a seafood paella on the quayside at Puerto Banus—or next time their politicians get pushed around by the EU in international fisheries negotiations. As the most powerful fishing nation, Spain tends to call the shots in Europe.

The expansion of the Spanish fleet coincides exactly with the time known euphemistically to modern Spaniards as the "years of isolation" and to the rest of us as the era of General Franco. I am indebted to a paper called "Spain and the Sea" by two academics from Seville University for explaining the connection. It points out that Spain nourished grandiose overseas ambitions and stimulated shipbuilding in the early twentieth century, but it was under General Franco's fascist government, which looked back approvingly to that time, that the fishing fleet underwent enormous growth. Fishing was the beneficiary of Franco's policy of protection and intervention, in which he favored the navy, the merchant navy, and shipbuilding. Spain's fishing fleet, in poor shape at the end of the Spanish Civil War, began to grow, and so did its catches—from 440,000 tons in 1940, the year after the Spanish Civil War, to 1,647,854 tons (a record never since surpassed) in 1974, the year before Franco's death. As a Spanish member of Greenpeace was the first to point out to me, joining the EU meant that the size of Spain's fishing fleet was officially supposed to be capped—one of several rules Spain's fishermen have been struggling against ever since. So it can be said that General Franco still has an outsized influence on Europe's fishing politics. No doubt he would approve.

Rules exist that should ensure most seas are managed commons. In fact, lack of enforcement and too many fishermen—Garrett Hardin's two definitions of management failure (see Chapter 9)—mean that Europe's seas are unmanaged commons in which the commoners and their fishy livestock face inevitable ruin. The growing contrast, I am beginning to think, is between those areas of the world where fisheries are commons, and those parts of the world where they are not.

Peterhead fish market, Scotland. 7:00 a.m., January 1997. The year Tony Blair became prime minister of Britain. Six thousand boxes of cod, haddock, monkfish, turbot, halibut, ling, and saithe were being sold. I was there to investigate the extent of "black fish" landings at a time when nature had just delivered one last, miraculously large year-class of cod in the North Sea.

The port was promoting a new policy of high prices for

fewer, high-quality fish, but this wasn't working. The harbor-master had asked the buyers and auctioneers, yet again, to wear white coats and not to walk on the boxes of fish. The buyers clustered around the auctioneer, and when they could not catch his eye, they clambered on the fish.

Slowly it became clear what we were seeing. The black fish were right before our eyes. Landings are measured in boxes, which at Peterhead should contain 125 pounds of fish, but all of them were overfilled. They won't stack readily without crushing the fish, so they must be reloaded by the buyer. An overfilled box, rather like a baker's dozen, theoretically meant a better price, and the extra volume went unrecorded because the boxes were not weighed.

Low prices were another indication of something afoot. It was the end of the month, so quotas should have run out and fish supplies should have begun to dry up, but prices were flat. People admitted, quite openly, that truckloads of fish were landed in small ports around Scotland and driven to processing factories or wholesale markets in Hull or Grimsby. A "private" sale, as it is euphemistically known, was agreed upon while the boat was still at sea. Fishermen would tell their friends where the inspectors were. Landing a few hundred boxes takes only a few minutes with a fork-lift.

"If the inspector should show up, the fishermen will simply declare the boxes he is landing," one person in the industry told me. "If the inspector does not appear, the skipper will unload his overquota fish, then take the rest to one of the larger ports, such as Peterhead, and offload his legal catch." Over time the skipper will keep spare quota, thanks to underreporting, so that he can go on fishing when, in fact, his vessel should be back in port. That week I was in Scotland there were four landings in the tiny port of Montrose—"that well-known fishing port," as one harbormaster observed dryly.

The fisheries inspectorate was poorly paid, subject to intimidation, and had very little political support. When I was at Peterhead five years earlier, people used to deny there were black fish landings. But not now. "There are very few people clean on the catching side in the industry now. Illegality has been institutionalized," a board member of the government's Sea Fish Industry Authority told me.

The Scottish Fishermen's Federation admitted that their members were landing black fish. Bob Allen, chief executive at the time, explained that this was just the "real world" his members lived in. One Scottish fisherman explained it simply: the cut in quotas that EU ministers had agreed upon made it necessary to break the law to stay in business. "The way we have been regulated means that everything we were doing is now illegal. The ministry knew this would happen. My boat was built with an EU subsidy and was refitted with a ministry grant. We've got mortgages to pay. It's like telling British Airways that they can only fly to Paris once a day when they have been used to flying six times."

A senior inspector then told me that 50 percent of the cod and saithe landed was illegal. Lord Selborne, chairman of a House of Lords select committee, said it was 40 percent. Whatever the figure, the scale of the landings overwhelmed the 12 percent cuts in fishing quotas for cod agreed that year. Over the next couple of years a huge year-class of juvenile cod was squandered, mostly as bycatch.

Meanwhile, the black fish were on the truck and heading south to where the processors are, in Hull or Grimsby, which were the home ports for the almost defunct long-distance fleet.

Documentation was, and still is, easy to falsify. If a truck is stopped, there is no way of telling where the fish are from, and the inspectorate does not consider it cost-effective to monitor the fish after the first sale.

Large companies involved in fish processing sell to the supermarkets, so the fish one eats, if it comes from the North Sea, stands a 50-50 chance of being black. These big companies have shareholders who are not used to dealing in illegally obtained goods, so questions have been asked. The directors of Booker Fish, the largest fish processor in Britain, sought legal advice on whether it was an offense, knowingly or otherwise, to buy illegally caught fish. Their lawyers said that the offense was catching it, not buying it, unless collusion could be proved.

Another processor said: "You get a call saying, 'Do you want 200 boxes of haddock?' It arrives by truck. It goes into the major processors. They won't know that it's illegally caught. They won't ask where it is from. They'll ask what the quality is and when it's going to arrive."

Some of the more upstanding firms could not cope with the fiddling that erupted across the industry when the quotas, aimed at saving fish for the future, began to bite. Irvines of Aberdeen told *Fishing News* that black fish were part of the reason for their decision to quit the business. Some large firms had been told by their accountants that they were not viable without black landings. Half the fishermen in Peterhead are teetotaling Presbyterians; the saying goes that the other half drink twice as much to make up for them. As one Grimsby merchant put it: "I've got friends in Peterhead who wouldn't buy fish caught on a Sunday. But they have a responsibility to their staff and they have had to become complicit with black fish." Very few would have nothing to do with it. Only one processor on the docks at

Peterhead, a member of the Free Church of Scotland, was said to refuse to handle black fish. The understanding at that time was that nearly everyone else was willing to.

Although fish quotas were being cut, boatbuilding on the east of Scotland was booming. Boats theoretically designed to catch deepwater species to the west of Scotland were being turned out of Scottish yards. What they were actually going to catch was black fish. Boats were being built with little quota to fish on. Bank managers were aware of this quandary. One asked a skipper just how he intended to finance his boat from his modest quota and was told it would be from black fish.

David John Forman was chief executive of the largest fish processing company in Peterhead. Outside his office window an army of women in sterilized overalls and hats were filleting fish. A newspaper clipping on his wall recalled how he pulled the crew of a Danish trawler from their sinking vessel in a force ten gale. He recalled the moment vividly and described how, as he arrived at the wreck, the lights of the trawler were already shining under the sea.

I asked him about black fish. He replied: "The quotas are uneconomic." His meaning was plain: skippers would go bust if they stuck to their quotas—and nobody was going bust. In his view, "the government has been quite tolerant."

It was, after all, an election year.

That was nine years ago. When I published most of the above in the *Daily Telegraph*, there was an outcry. Many British people had been brought up on the great myth of the commons, that only foreigners cheat. The news hit their ideas hard—for a day or two. There was a crackdown of sorts by the newly installed Labour government, which insisted that fishermen land at designated ports, as they did in Holland. My contacts told me not to come to Peterhead or I would be beaten up.

Tony Blair is still in power as this book goes to press. There are many fewer cod in the North Sea, and every other cod landed is illegal. About a thousand fishermen left the industry last year. The trail of slime from black fish continues to touch all levels in the fishing industry, together with the banks who lend to fishermen, the food processors, the supermarkets, and even the consumer, who benefits from continually flat prices.

A significant change that *has* come about since 1997 is Scottish devolution. Scotland now has its own Parliament with competence for devolved matters including agriculture and fisheries. This has made Scottish politics even more like those in Spain and Newfoundland. Little time in the Scottish Parliament is spent inquiring what is best for the fish and more is spent inquiring why scientists have not allowed fishermen enough quota, in any given year, to make a living. The Scottish media report sensitively on the plight of the fishermen and what politicians of various parties say they are going to do about it, while London-based or "English" media are currently more likely to report on fishing as a story with two sides about the degradation of the environment and the slow death of an industry.

An obligatory ingredient of any newspaper story about fishing in Scotland seems to be a quote from the Cod Crusaders, two fishermen's wives from Fraserburgh, Carol MacDonald and Morag Ritchie, who have the slogan "Save our cod."

This does not mean what you might think it means. In a telling exchange before a House of Lords select committee, Lord Selborne asked Elliot Morley, the London-based fisheries minister: "So save our cod means kill our cod?"

"Yes," replied Morley.

In its legal letter of complaint to the United Kingdom, once a more law-abiding nation, the European Commission noted a failure by the authorities to cross-check data and a failure to take action against fishermen who broke the rules on a number of occasions. The commission noted cases of misreporting, where vessels recorded catching fish in one area while their spy-in-the-sky satellite records showed that they were in another. To avoid detection, vessels often turned off their blue boxes altogether. This did not seem to be reported or penalized by the authorities, who, despite much fanfare in the 1990s when they were introduced, have yet to use the blue box as an enforcement tool.

Fishermen were reported using tricks which the British once thought of as Spanish: concealing cod in large containers labeled ling, and misreporting saithe, cod, hake, megrim, and monkfish as ling, greater forkbeard, tusk, and dogfish.

Blair, who had done almost nothing in the previous seven years for the fish stocks around Britain's shores except eating fish in preference to beef at official functions, set up an inquiry in 2003 into the fishing industry. Written by the prime minister's Strategy Unit, a think tank run by some of the brightest civil servants, the inquiry's report turned out to be a thoughtful and comprehensive piece of work that told the industry to reform or die, albeit in relentlessly upbeat, positive terms. It identified lack of compliance and the excessive size of the white fish (cod, haddock, and plaice) fleet as the "two major challenges to sustainability and profitability." The changes it recommended to stop black fish landings—administrative penalties, such as the loss of a license for infringements, electronic logbooks linked to markets, onboard observers, and improvements to the onshore paper trail—were sensible enough. Only the last has been adopted at the time of writing.

Illegal fishing is not unique to Britain and Spain. It exists in all jurisdictions, though in few at such a stupendous level. Yet wherever illegal fishing challenges the survival of species of fish, or the dolphins, turtles, or seabirds that are the bycatch, the whole food industry—the sellers and the buyers in supermarkets and processing firms as well as the fishermen—bears some responsibility for what has been going on in the sea. Given the scale of illegal catches globally, it is virtually impossible for anyone to put their hand on their heart and say they have not inadvertently bought, sold, or eaten illegally caught fish. No one can back away from that, not the humblest fish shack nor the most famous restaurant in the land.

11

DINING WITH THE BIG FISH

Once there were some simple certainties about fish. We believed fish was largely fat-free protein and therefore better for you than hamburgers. We believed that the nations who ate fish as a large proportion of their diet, such as Japan, had far lower levels of coronary heart disease than the west. We believed there were plenty more fish in the sea. Since then, things have gotten a whole lot more confusing. The reason, partly, is the extraordinary public debate in the United States between scientists who believe that eating some large predatory fish can give you long-term mercury poisoning and others who believe that the omega-3 fatty acids contained in oily fish provide such a wide range of health benefits that we should go on eating some anyway. When people who care about food then try to factor in concerns about overfishing, they are plunged into almost comical decision-making paralysis whenever confronted with a restaurant menu.

Perhaps it is because I live in Europe—where the health advice is the same, but mercury is barely an issue—that it seems to me there is a perfectly commonsensical path to tread through the health-scare minefields without any need to panic along the way. Small children and women of childbearing age should avoid eating large predatory fish that concentrate mercury—or at least avoid eating large amounts—and anyone else concerned about mercury in large fish should either eat them in moderation or take their omega-3s from low-mercury fish lower down the food chain (wild Pacific salmon, mackerel, herring). Fish, eaten with an awareness of the latest health advice, is still healthier than hamburgers. And we'd still be healthier if we ate as much of it as the Japanese. But if we did there would be even less fish left than there is already.

Americans eat remarkably little seafood, on average, compared to other nations: almost 17 pounds of seafood per head each year, while in Britain people eat 44 pounds, in Canada 52 pounds, in Spain an impressive 97 pounds, and in Japan an imperial 128 pounds. Which makes it all the more strange that the U.S. media have such a fixation on which fish are safe to eat. Consumption, though, is rising, despite a slide in the purchasing of tuna after the official FDA advice on mercury in fish was published in 2004. What frustrates me is that while the tuna industry and its supporters, on one side, and the Food and Drug Administration, on the other, duke it out over the relative merits of oily fish, some of the wider questions about what we want from our food and from the sea go unasked.

We have seen that the world's wild fish stocks have peaked and are now in decline. Yet no one seems to consider whether there are any human health implications to this. Needless to say, there are if you happen to starve because the fish you depend on have run out—and it is a real question how much longer some poor nations that depend on fish will be able to go on depending on them.

For those of us who live in the countries that import those

fish, there are more questions. Fish are the last foodstuff, for most of us, that comes straight from the wild. Fish range extensively, and this partly explains the flavor, the firm texture, and the healthful properties of a cod steak or a plate of sardines grilled Mediterranean-style. If a sustained attempt is not made to save the most popular species of wild fish, we are going to be left mostly with farmed ones. Are we happy about that?

Farmed fish have the same problems amplified by unnatural confinement that intensively bred livestock do. They need drugs to treat illnesses and pesticides to kill parasites. They lead brief, sedentary lives like other domesticated creatures. They are fed with ground-up smaller fish, which are often themselves overfished. This concentrates the pollution present in the sea in the fatty tissues of the farmed fish (as eating smaller fish does naturally in the flesh of larger, longer-lived predatory fish). In short, farmed fish have all the problems that have led people to stampede away from intensively farmed meat and toward locally grown or organically produced livestock. For this reason the conservation of wild fish is a human health issue as well as an environmental one.

Given the choice, most of us would prefer to eat wild fish. So it is rational to consume wild fish in a way that promotes their continued abundance. It makes sense for consumers who want the species of wild fish that currently exist to be available in the future to avoid eating endangered species and to favor other fish that are less wastefully caught, using more selective methods. Indeed, many consumers already take the trouble to choose fish ethically according to what information is available; several environmental organizations now provide user-friendly seafood guides for conscientious eaters. Which makes it all the more surprising that many of the professional buyers, preparers, packagers, and servers of fish still aren't making the connections. For instance, in some of the world's celebrity restaurants—in which the chefs are as much celebrities as the clientele—the marine equivalents of the panda, the rhino, and the great apes are on the menu.

Nobu Matsuhisa stands out as a brilliantly inventive exponent of Japanese and fusion cooking and an ambassador of Japanese culinary tastes and techniques to the world. His thirteen Nobu restaurants are the haunts of film stars, supermodels, and tennis stars on three continents. Nobu was trained in Japan and cooked in Peru, Argentina, and Alaska before opening a restaurant in Beverly Hills into which Robert De Niro, actor and part-time restaurant entrepreneur, one day happened to walk. The friendship and business partnership they formed is one of the greatest success stories in the restaurant business. Nobu is known for creating imaginative, expensive, and stunningly beautiful dishes that fuse the pure, bland tastes of sushi and traditional Japanese cooking with the North and South American flavors of garlic, chili, and coriander. He is perhaps even better known for the clientele of his restaurants, where the reservations can read like a night at the Oscars: Gwyneth Paltrow, Renée Zellweger, Nicole Kidman, George Clooney.

Nobu has become a master of publicity as well as of sushi and sashimi, and he has used this mastery to expand the market for his kind of Japanese food. He teamed up, for example, with Martha Stewart to make a Webcast in the Tsukiji fish market in Tokyo, enthusing over all those beautiful tuna, before Stewart mounted her own invasion of the Japanese market with her particular brand of cooking and decorating accessories. His first cookbook, *Nobu: The Cookbook*, had more celebrity endorsements than any book, let alone a cookbook, you have ever read.

Words of praise from Madonna, Giorgio Armani, Bill Clinton, Andre Agassi, Robin Williams, Cindy Crawford, Leonardo DiCaprio, and others jostled for space on the back cover. Kate Winslet, for instance, said: "The food at Nobu can be described in two ways—quite simply heaven on earth and SEX on a plate." Stephen Spielberg said: "Your food is the BEST. Just don't tell my mother."

The list of fish Nobu chooses to cook may be sublime in terms of quality, exclusivity, and taste, but if you know your ecology, you would have to say he is vulnerable to criticism. Check out the recipes in his cookbook against the prevailing ecological wisdom on what not to eat, set out, for example, on the Seafood Watch Program Web site run by the Monterey Bay Aquarium in California, the guide published by Blue Ocean Institute, or the National Audubon Society's Seafood Guide. In Nobu's cookbook you will find many species that are in very poor shape, such as abalone, Caspian caviar, Chilean sea bass, grouper (a coral reef fish), red snapper (occasionally fished with dynamite), sole, and the finest sashimi-quality tuna steaks.

In this book Nobu makes no reference to which tuna he uses, describing all tuna as simply *toro*. He does, however, say he prides himself on the very finest-quality ingredients, so the culinary world would immediately assume he means bluefin or bigeye, the most sought-after tuna. I contacted his PR people in New York but they declined to provide this information. Nobu's cookbook also includes exotics such as flying fish roe (which does make you wonder what happens to the rest of the flying fish) and the notorious blowfish, or fugu. Now, the fugu can kill *you*, which is something Nobu obligingly draws attention to, but he never mentions anything about the lethalness to fish species of any fishing methods or management regime. To be fair, it has been the convention not to talk of such matters as sustainability in cookbooks for fish. One wonders whether that convention is itself sustainable in the reduced circumstances in which the world's fish stocks find themselves. Given Nobu's elevated status among chefs and his status as a role model among those who follow his fusion of Asian and Western cuisines, that omission is a shame.

On the West Coast of the United States, where the "Take a pass on Chilean sea bass" campaign altered the menus in many restaurants, publishing recipes for a fish as controversial as this, as Nobu does, would appear to be a statement of defiance. There are, of course, some complex ethical issues about Chilean sea bass (or Patagonian toothfish, as it is known elsewhere), which environmentalists, in their hurry to project simple messages, don't often tell you, and which Nobu might have taken the opportunity to mention in justification of his decision to serve it. A simpleminded boycott, as advocated by U.S. environmental groups, devalues not only the fish caught by transoceanic poaching syndicates but also the legally caught Australian toothfish and South Georgia toothfish, which are sustainably managed. But how do you know the fish on your plate is legal or sustainably caught? Nobu doesn't say where his toothfish comes from. I addressed that question to his publicist in New York and also asked about his fish purchasing policy in general. She called back to say Nobu had "no comment on the issue." So without proof of any chain of custody to identify legally caught or sustainably caught toothfish, we are unable to assume that Nobu knows where his toothfish come from. His Chilean sea bass and truffles with yuzu soy butter sauce may, in other words, just be plain old poached toothfish.

Does this matter? Yes, I think it does, because the attitudes of the great are the stuff of fashion. Celebrity chefs are the leaders in the field of food, and we are the led. Why should the leaders of chemical businesses be held responsible for polluting the marine environment with a few grams of effluent, which is sublethal to marine species, while celebrity chefs are turning out dead endangered fish at several dozen tables a night without enduring a syllable of criticism?

Nobu certainly isn't the only celebrity chef who doesn't tell you where his fish come from—this problem pervades the high-end restaurant industry. When I was in New York last year I noticed that one of the L.A. restaurants in the Nobu mold, called Koi, had opened a branch in the Bryant Park Hotel. Koi retails contemporary Japanese food to New York's rag trade district, mixing traditional and Western ingredients. Koi's menu is big on imported shrimp, which the National Audubon Society says you should avoid. Its chef, Sal Sprufero, has a signature dish of—you guessed it—steamed Chilean sea bass and shiitake mushrooms. Perhaps Sprufero has found a way to get this from one of the sustainable sources in South Georgia or Australia. All credit to him if he has—but his publicist declined to answer any of the questions I sent her. What gets me more worried is that the sample sushi and sashimi menu on his Web site is headed by bluefin tuna.

Nobody, I believe, should be knocking out a dozen servings a night of bluefin tuna, which is already in the anteroom to extinction. Least of all should anyone be treating the small and dwindling western Atlantic bluefin population as an unlimited resource. To put it another way, how would you feel about a slice or two of mature endangered orangutan, served raw? If high-end restaurants did that I believe most movie stars would walk out rather than pose with the chef.

You'd be amazed by the number of chefs who serve bluefin in New York City. There are high-end Asian eateries, such as Megu, a Japanese joint with towering prices and a baffling thirteen-page menu. There is Jimmy Sakatos, chef at the Carlyle, a hotel favored by political dignitaries, celebrities, glitterati, and in-the-know-travelers, according to its literature. And then there is Laurent Tourondel, one of the top-bracket chefs, whose swanky BLT Fish offers a selection of endangered and overfished species including Atlantic bluefin tuna (International Union for the Conservation of Nature listing: critically endangered), Icelandic halibut (IUCN listing: vulnerable), South Pacific swordfish (not a bad choice, sustainability-wise, but watch the mercury), and Florida red snapper, a reef fish that lives up to fifty years of age (though it is usually caught at the age of three to four). Red snapper is the fish with the longest record of overfishing in the Gulf of Mexico.

David Bouley was for many years New York's favorite chef, opening Montrachet, which rated three stars, in Tribeca in 1985 and the reliable Bouley in 1987. The latest version of Bouley has on its menu seared swordfish (though it doesn't say where these are from), Maine baby skate (can't he wait until they have spawned?), and Nova Scotia halibut (IUCN status: vulnerable). However much you admire Bouley for turning one of his restaurants into a canteen for Red Cross workers at ground zero in the wake of the Twin Towers atrocity, you do wonder if he has had time to catch up on other aspects of the news.

These are just a few examples, of course, of what's on the menus in high-end restaurants throughout New York City and around the country. Their chefs, holders of star ratings from the *New York Times*, Zagat, or the Michelin Guide, feel themselves to be caught in a quandary, for a certain kind of customer at the world's top restaurants has always wanted something no one else can have. To some diners, rarity is just another form of exclusivity. The chef's fear, apparently, is that if some customers don't get that exclusivity, they will go somewhere else. On the other hand, you might think that responsible chefs could win new customers in an era of rising concern about declining fisheries by making what *is* plentiful more appetizing.

Eric Ripert, executive chef of Le Bernardin, provides a heartening example that other chefs should follow. Ripert is arguably the best chef in New York, judged by the four stars his restaurant was given by the *New York Times*. He has subtly updated himself and responded to new trends. Le Bernardin bucks the globally declining status of French eateries and the attitude of many French chefs to endangered species of fish. He appears to believe that sustainability is part of the overall quality standard the top eateries should be hitting. I think he is on to something.

Remember, Ripert runs a fish restaurant, with a whole menu based on fish. In the present state of world fish stocks it almost impossible to run a high-end fish restaurant without serving something that is arguably overfished, but there is something rather admirable about trying to do the right thing in almost impossible circumstances. Ripert gives swordfish a break by not serving them, and refuses to serve Chilean sea bass either. In interviews he talks up the need for more organic farmed fish (though he prefers wild, having reservations about fish farming). He joined the Seafood Choices Alliance, the environmental network concerned about overfishing, in 2001. Seafood Choices tell me they operate a "big-tent" approach. Even if chefs aren't willing to change their menus to all sustainable stocks immediately (assuming this is possible), the Alliance welcomes those who take baby steps along the way. It believes that in the present state of the world's fisheries it is better to start somewhere.

So it seems churlish to point out that Eric Ripert also, at the time of writing, serves Iranian osetra caviar—the eggs of a sturgeon that is listed as "endangered" by IUCN in all the areas it is found: the Caspian Sea, the Black Sea, and the Sea of Azov. Endangered according to IUCN's definition means "facing a very high risk of extinction in the wild in the near future." Caviar is something that major-league chefs seem to feel their customers must have, but there is a problem: every single kind of sturgeon that caviar comes from is endangered.

Osetra caviar comes from what is known as either the Persian sturgeon or the Russian sturgeon—same fish—and is native to the southern part of the Caspian Sea and the Black Sea. It is smaller than the beluga, the largest and most endangered sturgeon that migrates from the Caspian Sea up the Russian and Kazakh rivers. Its caviar has a distinctive nutty taste which made it the favorite of Ian Fleming, the creator of James Bond. All Caspian sturgeons have been subject to an epidemic of poaching by Mafia-like gangs since the collapse of the Soviet Union, though the situation is said by some to be slightly more under control in Iran. In response to the poaching epidemic, the United States introduced a ban on importing beluga caviar in autumn 2005. CITES followed the United States's unilateral action in early 2006 by banning virtually all international trade in Caspian and Black Sea caviar. In spring 2006, CITES appeared to concede a difference between the desperate plight of other sturgeons and the Persian sturgeon when it decided to allow the export of Iranian osetra caviar for the rest of 2006. No other sturgeon could be traded globally that year because the Caspian and Black Sea states still had not agreed on a conservation plan. But the Persian sturgeon, though still endangered, is once again being traded.

Ripert was one of the few chefs to call me back himself. He told me that he had been consulting on the question of whether osetra caviar came from sustainable sources, as he was trying not to serve any endangered species. He had been assured by a source at the Blue Ocean Institute that the Iranian osetra trade was "extremely well regulated"—unlike that in Russia. All the same, he had decided to switch slowly to "Calvisius" farmed caviar, which in his opinion beat all other farmed caviars on taste. Calvisius caviar is from white sturgeon farmed in the hot water from steel mills in the Northern Italian town of Brescia.

Daniel Boulud, another French chef of renown, was still offering Azerbaijan (now banned from importation) and Iranian osetra caviar on the menu of Daniel, one of his New York restaurants, when I last looked some time after the CITES ban in early 2006, along with other fish from the red (don't eat) section of the good fish guides: grouper, cod, and Dover sole. Caviar that has already been imported may still be sold legally, but I felt there were questions I would like to ask Boulud about what he was going to do next in the light of the worsening plight of sturgeon. He offers something called "private stock" caviars on the Internet. Private stock caviars are relabeled, replacing the exporter's brand with the seller's own label, a practice that is perfectly legal but some might feel adds unnecessary complexity to the clear chain of custody expected under a treaty regulating international trade in endangered species.

I asked Boulud's people whether he had a sourcing policy for the fish he serves. Did he reject any fish from his restaurants on sustainability grounds? What was his policy on caviar, particularly after international trade was banned in early 2006? I got the following response from Georgette Farkas Trapp, director of public relations for the Dinex Group, Boulud's company: "We are happy to say that we do our best to remain informed of environmental concerns in the seafood market and to direct our purchasing accordingly. We have longstanding relationships built on trust with reputable seafood suppliers. We have faith in their knowledge and professionalism and depend upon them for their guidance in our purchase of sustainable fish and caviar." In other words: *We have decided to make this someone else's problem.*

While the caviar trade ban continues, there is a viable domestic alternative: caviar from farmed American sturgeons and paddlefish. The white sturgeon, an American species, produces a caviar very similar to osetra. Some of the companies selling it, and many West Coast and Las Vegas chefs, call this Californian osetra, which is potentially confusing and bolsters a market for newly imported osetra that many environmental groups feel is unsustainable. Rick Moonen, a New Yorker who has now moved his Restaurant RM to Las Vegas, is a supporter of the organization Caviar Emptor and a tireless advocate of farmed American caviars. He says white sturgeon caviar is just as good as osetra. When he was in New York he proved it to the satisfaction of those present by serving an American caviar sampler with farmed white sturgeon, farmed paddlefish, and farmed rainbow trout caviars.

There is beginning to exist at the top of the culinary profession a group of concerned chefs, including Moonen and Ripert, who see the sustainability as an integral part of the overall quality of the food they serve. Peter Hoffman is owner of Savoy, a Soho restaurant serving Mediterranean-inspired food. He is also chairman of Chefs Collaborative, an alliance of chefs who try to choose more sustainable sources of fish. Hoffman took Caspian caviar off the menu a long time ago. He lists Traci Des Jardins of Jardinière in San Francisco, and Paul Wade, now of the Four Seasons hotel in Houston, as other chefs outspoken on the merits of American farmed caviars.

There are ethical quandaries much closer to home than the Caspian Sea. One is the New England cod. You won't find cod on the menu at Savoy. You won't even find environmentally friendly hook-caught cod from the innovative Cape Cod Commercial Hook Fishermen's Association, which is endorsed by the Chefs Collaborative. Hoffman abides by the official advice, which is that New England cod remains overfished. There are just too few cod facing too much fishing pressure.

Yet you will find cod on the menu all over New York and up and down the East Coast. You will find it on the menu at the Legal Sea Foods chain, once described as one of ten classic American restaurants by *Bon Appetit* magazine. Roger Berkowitz, CEO of Legal Sea Foods, has said in interviews that he will not serve Chilean sea bass or orange roughy because they are overfished. Nor will he buy swordfish pups (fish of 100 pounds or less) because they haven't reached sexual maturity. Good for him. But it seemed to me, reading the seemingly endless preamble about quality with which the *New Legal Sea Foods Cookbook* opens, that this wasn't good enough.

Legal Sea Foods—slogan: "If it ain't fresh, it ain't Legal"—got its intriguing name from Berkowitz's grandfather's meat and produce store in Cambridge, Massachusetts, which was called Legal Cash Market, after the Legal trading stamps that were handed out to buy customer loyalty. His father named his new fish store Legal Sea Foods. Grandfather and father

Berkowitz established the family businesses' reputation for stocking only the freshest, highest-quality meat and fish. In his book, Roger explains how his father went to great lengths to find new ways to purify shellfish and to buy fish from the "top of the catch"—the fish caught last before the boat returned to port. And then he goes arguably too far. He writes: "My motto might be a sign I've hung on my office wall that says, 'If you re-fuse to accept anything but the very best, you will very often get it.' "

So if the best cod is either hook-caught cod, or no cod at all, according to the Chefs Collaborative, where does that put Berkowitz? I called him, and he told me: "Cod is one of the great issues in the company at the moment. The conclusion we are reaching is that we will not use cod in a multiplicity of dishes like we did before." He says they'll list one loin of cod on the menu, sourced from hook boats in Gloucester or Chatham, and that's it. He says it's a balancing act, supporting the industry but trying to back day boats and sustainable fishing methods. Well, he's ahead of the field in thinking about it. And so he should be.

We can't let the clientele of the world's pricey restaurants off the hook, either. It's they who are paying, and they who call the tune. I'm curious. Do the likes of Gwyneth Paltrow, Chris Mar-tin, and Bill Clinton never ask the waiter where their fish is from and how it is caught? After all, I have seen Gwyneth posing in a photograph with Nobu, and her British husband evidently has a political conscience, as he promotes fair trade products relent-lessly.

Do the fashionable clientele of top-rated restaurants believe that the international standing of the chef whose food they are eating somehow means it must be all right? Or does their own desire to eat healthily simply come before asking difficult questions? You can cruise the world's millions of omega-3 Web sites without encountering any reflections about where these prized fatty acids are coming from and at what social or environmental cost. For some people, what goes into their bodies has become an overriding obsession. Perhaps we are witnessing a successor to the Me Generation—namely, the Don't Care About the Rest of the World as Long as I Have a Spa and Some Omega-3 Fatty Acids Generation. Let's call it the Omega-3 Generation for short. Or is that thought just too depressing?

I prefer to believe that the canny, Toyota Prius-driving stars we are talking about just don't know much about fish. Having learned a little more, they will ask whether there's any ecofriendly stuff on the menu, and if not, they'll be out of that trendy restaurant faster than you can say "politically correct." After all, the time is surely coming when being caught eating endangered species will be a whole lot worse than being seen in real fur.

DEATH IN A CAN

We can all make fun of the superstar chefs and the celebs being ushered to their reserved seats, but we ordinary mortals are implicated in all this, too. Before we get too smug, we need to think about the fish most of us eat when it is not our big night out—the can of tuna in the pantry or the sandwich filling in the deli. Few pantries in the developed world do not have a couple of cans of tuna in them, waiting to be used in a quick casserole or salad. As the white fish stocks of the Atlantic have waned, and trade in fish from oceans further away has made up the shortfall, the trend has been toward Mediterranean- and Japanese-style food, even in countries that used to eat something else. In Europe the spread of Mediterranean eating habits—for health reasons, too—has meant an upward surge in tuna sales, to the point where the EU is now the biggest market in the world for sales of canned tuna, just edging out the United States.

This might be because tuna has had bad press lately in the United States. In early 2006, an FAO tuna market report was predicting that the mercury issue might yet have a disastrous impact on the U.S. industry. The *Chicago Tribune* wrote in early 2006 that U.S. Food and Drug Administration tests showed some light tuna was high in mercury. (Previously the assumption was that mercury was exclusively in dark tuna.) On the other hand, the industry, represented by a lobby group called the U.S. Tuna Foundation, said that canned light tuna was well below the FDA limit and the new figures actually showed how safe canned tuna was. The majority of canned light tuna is the skipjack variety, which grows very quickly to a modest size and does not live long enough to accumulate mercury compared to the longer-lived yellowfin. The FDA appears not to have realized previously that some light tuna is longer-lived yellowfin. The damage to the market was done.

I've already explained where I stand on mercury—heavy metals are nasty stuff, usually found in very small quantities, but they do accumulate in the body, so take the health advice and either avoid those varieties of fish or eat them in moderation, or eat fish low in mercury and high in omega-3s, of which there is plenty. What interested me, rather, in this discussion was why the terms "light tuna" and "dark tuna" were used at all. Why did canners not say from the start what species was in the can? For that matter, why did they not tell us a whole lot more about where the tuna is from, how it is caught, and whether it comes from overexploited stocks? I bought a can of tuna at random from each of the three major brands at a supermarket in New York and studied their labels. The Starkist can said it contained a light tuna fillet. It did not tell you which species it came from or even the ocean in which it was caught. It did tell you a whole lot about total fat, saturated fat, fiber, cholesterol, sugar, protein, and vitamins, as required by the FDA. Oh, and it told you it was "dolphin-safe." The cans from Bumble Bee and Chicken of the Sea said they were chunks of albacore and dolphin-safe, but again, neither even mentioned which ocean the tuna was caught in. I found this curious. When I got home, I checked the store brands I found on my pantry shelves in England. The EU, I discovered, doesn't require canners to tell you what fish you are eating, either, but it does at least require the canners to say which ocean it is from.

It strikes me that one of the ways of feeling less concerned about one of your fellow creatures is not to give it a name. Not telling you even which ocean it is from means you can't match up anything you happen to know about tuna fisheries with what you are eating. Tuna, more than any other fish, is a global commodity. The cans on my shelf in Europe were from Mauritius, Thailand, the Maldives, and Ecuador. The tuna industry in the United States also imports tropical tunas from all the oceans of the world. Almost certainly the reason for the reluctance to provide information on country of origin was originally practical: companies often switch from one supplier to another so including too much geographical information would mean canners had to print more labels and spend more time making sure they went on the right cans. But as we learn more about the crisis in world fisheries, the continued anonymity given to so much tuna invites suspicion.

What I found particularly surprising was that although half the tuna canners didn't bother to tell you what kind of fish was in the can, the one piece of ecological information nearly all want to tell you is that their tuna is "dolphin-safe" or "dolphin-friendly." So what does "dolphin-friendly" mean and why should it be so important to fish canners that they would go out of their way to tell us about it? I remembered that dolphin bycatch was a problem for the catchers of yellowfin in the eastern Pacific, but I had read that there were also other bycatches, such as turtles and seabirds. So, I wondered, why didn't they tell us it was turtle-friendly tuna, or seabird-friendly tuna, or shark-friendly tuna, or even tuna-friendly tuna (caught in a way that means there will always be more tuna)? The reason for not telling us that, I have now discovered, is probably because the way they catch the tuna that goes into cans is not friendly to very much at all.

Tuna canned for the United States and Europe is mostly caught by purse seiners. The FAO estimates the size of this fleet at 570 vessels worldwide, ranging from 275 to 4,400 tons. Spain and France are big tuna-fishing countries, especially in the Indian Ocean, the Atlantic, and the Pacific. The United States is a big importer from countries including Mexico, Panama, Venezuela, and Ecuador in the eastern Pacific. The major catching nations in the west and central Pacific are the Japan, Taiwan, the United States, and Korea. Australia is a big catcher in the Pacific and the Indian Ocean.

Tuna eaten fresh as steaks, sushi, or sashimi comes from the long-line fleets of countries such as Japan and Taiwan, whose baited hooks and lines up to 80 miles long cause a significant by-catch problem for albatrosses, endangered turtles, and sharks. The UN FAO has estimated that there are approximately 1,556 long-line vessels of larger than 110 tons with freezing capacity catching tunas. The other significant way of catching tuna, particularly around the islands of the Indian Ocean, is by using pole and line from bait boats. Bait boats attempt to create a feeding frenzy by throwing chopped-up bait at swimming schools of small tuna, often using fire hoses to whip up the surface water. Then individual tuna are hauled out by pole and line. The size of the tuna caught in this way tends to be small, but there is virtually no bycatch. The resulting tuna supplies the growing fresh tuna market. This pole-and-line method, also used to catch albacore in the Atlantic, is the least-bad method of tuna fishing.

The targets of the world's purse-seine fleet are the tropical skipjack and yellowfin tunas. The skipjack grows up to 3 feet long and is incredibly fecund, with a short life and a high natural mortality rate. It was described to me by the late Dr. Geoff Kirkwood of Imperial College, London, as the rat of the sea. I've also heard it described as the oceans' cockroach. Like mackerel, to which it is related, the stripy skipjack swims the seven seas, spawning all over the place, and is hard to fish out. "Nobody has yet managed to dent the population of skipjack. Indeed, I don't think anyone really knows what it is," Kirkwood told me. That, of course, doesn't mean no one will dent it someday.

The trouble begins, conservation-wise, because skipjack tend to run in shoals with the less fecund bigeye and yellowfin tunas. Yellowfin grows much larger than the skipjack, up to about five feet. It provides better-quality, firmer flesh used in sushi and steaks, and is also frozen and canned. It is fully exploited in most oceans and in some places is known to be exploited beyond its maximum sustainable yield. Then there is bigeye, a deepwater tuna that feeds in the cold depths and matures more slowly than the other tropical tunas but faster than the cold-water bluefins. It is prized for sushi, and the most valuable species after the bluefin. Adult bigeye tuna run much deeper than yellowfin or skipjack, at around 500 feet down. But juvenile bigeye and yellowfin tend to run with skipjack much closer to the surface.

There is now concern among several management bodies that bigeye is being overfished in large numbers in the Pacific, Atlantic, and Indian Oceans. The Red List, compiled by IUCN, lists bigeye as "vulnerable," which it says means "considered to be facing a high risk of extinction in the wild." Kirkwood said, "Both bigeye and yellowfin need to be managed," but he knew better than anyone that in the Indian Ocean this isn't happening. The scientific committee of the Indian Ocean Tuna Commission (IOTC), of which he was a member, recommends "a reduction in catches of bigeye from all gears [i.e., purse seiners, long-liners, and bait boats] . . . be started as soon as possible."

Concern arises from the way most purse seiners catch tuna—by setting nets around naturally floating objects or man-made FADs, a much easier way of catching tuna than setting on fast-swimming schools. A sleek, modern boat from either the French or the Spanish purse seine fleet picks up a signal from one of its FADs that has found the thermocline (where hot water meets cooler water) and begun to congregate tuna. It steams to intercept. It sends out a fast skiff, which drops off the net that duly encircles the shoal of fish around the FAD, and the mother ship begins to haul. Alternatively, the purse seiner may set its nets on whales—ironic, since the Indian Ocean is meant to be a whale sanctuary under international law—which usually escape by breaking through the nets, but sometimes don't. Setting on FADs tends to be highly indiscriminate, Kirkwood told me, in terms of the size of tuna killed and the number of other fish species caught.

Which brings us back to dolphins. They do not seem to be caught with tuna anywhere except in the eastern tropical Pacific, where the U.S. tuna fleet discovered in the late 1950s that the largest yellowfin tuna swim beneath dolphins. There was a massive bycatch of dolphins for three decades in the eastern Pacific after fishermen began to set their nets on pods of dolphins. In 1986 alone, some 132,000 dolphins were killed as bycatch in the eastern Pacific tuna fishery. This carnage understandably aroused the anger of environmental groups, in particular the Earth Island Institute (EII), an environmental group almost wholly focused on the impact of fishing on marine mammals. A series of consumer boycotts, lobbying activities, and laws in Congress ensued in an attempt to stop this unnecessary killing. However, the threat of the closure of the U.S. market to tuna from other countries, if it was caught in ways that killed dolphins, failed to produce the regional moratorium on killing dolphins that many conservationists wanted. A voluntary agreement (known as the La Jolla Agreement) devised by the Inter-American Tropical Tuna Commission, the regional management body, proved more successful. By allocating limits on the number of tuna each individual vessel was allowed to kill, the number killed declined sharply to less than 2,000 in all but one year after 1997. The agreement became a legally binding treaty, the Agreement on the International Dolphin Conservation Program, in 1999. In 2004 the number of dolphins killed by tuna fishermen setting their nets on dolphins—and attempting to allow dolphins to escape—was 1,461. The IATTC says that this is 0.015 percent of the total dolphin population and therefore sustainable. The Earth Island Institute still continues to attack the program and to urge its members not to buy "dolphin-friendly" tuna caught in this way, because of the numbers of dolphins killed.

That sounds reasonable, looked at from the dolphins' point of view. But the ecological death toll caused by the alternatives to setting nets on tunas—namely, setting on floating objects, free-swimming shoals, or FADs—had began to alarm other environmentalists. Fishing on FADs appeared to cause a bycatch of up to fifty times the number of other sea creatures to those caught in tuna sets, potentially including highly vulnerable species such as sharks and oceangoing turtles. As a result, environmental groups such as Greenpeace and WWF now support the IATTC's international dolphin conservation program, monitored by observers, which sets on dolphins but allows them to escape.

Differences of opinion remain. A conservation-minded tuna expert with a long time in the business told me: "Thanks to EII's extreme position, attempting to curtail all tuna fishing associated with dolphins, several countries like the U.S., Spain, Ecuador, and others started to fish with FADs, and today a lot of fishing is done on FADs, which means that instead of dolphins being killed, twenty different species are now in bad shape. They're catching fish that haven't had time to reproduce. From a disaster for dolphins, we have progressed to a whole ecosystem disaster."

"Dolphin-safe" remains the PR mantra of the tuna industry, even in oceans where dolphin bycatch has never been a problem. In other words, the market leaders in the canning business spend a lot of time claiming credit for not doing things in some oceans that have never been done. Whatever its intention, the result of this strategy has been to confuse the consumer about what was really going on by raising animal welfare concerns that were never really relevant, while failing to deal with genuine concerns about the fishing itself. To be fair, companies were responding originally to a kind of species-ism among certain animal welfare groups that appear not to see fish as animals at all, except as a commodity to be caught in an efficient way. To them I would say, look up *animals* in the dictionary. Fish are included.

The only reason there is not more of an outcry about what

goes on in places such as the Indian Ocean and the western Pacific is that nobody knows about it. Very little information exists about what is caught as bycatch or thrown away as discards. The reason is that hardly anyone has ever collected any. Why? I suspect because it has never been in anyone's interest to do so.

So in the Indian Ocean, where there has never been a problem of accidentally catching dolphins, what *have* they been catching accidentally? I spoke to Juan Morón, head of the Spanish tuna catchers organization in Madrid. He said that purse seining was a relatively clean operation compared to long lining. So, I said, show me the proof. He told me that some of the Spanish fleet had recently begun to carry observers, at the EU's insistence, as boats in the eastern Pacific have had to for the past fifteen years. Where was the data compiled by these observers from the Spanish Oceanographic Institute, which is notoriously secretive about its information? Morón said he'd try to get it for me. He came back and said the data had not been published.

I knew what to look for; I just wanted to see if anyone was interested in volunteering it. I tried the same question on Dr. David Ardill, then secretary to the Indian Ocean Tuna Commission (IOTC), based in the Seychelles. Under the heading "Discards" on his Web site, figures were said to be "available from the commission." When I asked to see them, nothing was forthcoming except an e-mail that explained there was little information about bycatch or discards, and what did exist was confidential. He explained: "There are enormous commercial interests involved, and the reality is that if we are not able to guarantee confidentiality, we would simply get falsified data, and would be much worse off than at present."

Surely he could let me see the information that was advertised as "available from the commission"? Apparently not. I wrote back that I thought his confidentiality rules—which, by the way, are to protect vessels that have found it more convenient for tax purposes to fly flags of convenience rather than the flags of their own countries—looked more like those of a pirates' club than an open and accountable public organization charged with making one of the world's great commodity foods sustainable. If he wanted people to go on buying cans of tuna from the sea he was supposedly "managing," he was going to have to do better than this.

All this time the big purse seine vessels, monitoring their hundreds of FADs and sonar buoys floating in the Madagascan current and scouring the sea with their long-range sonar, seemed to be getting away with their activities. Evidently, no one was going to invite me on board a Spanish or French purse seiner to count the bycatch. However, I did finally interview someone who had been on such a trip. My contact was Spanish and worked for an internationally known company that made specialist sonars for use in the tuna industry, one of the specialist companies' most lucrative markets.

My contact showed me photographs of FADs—square wooden frames with trailing tendrils of mesh hanging from them—being built by lean, fit men in dazzling sunshine on the deck of a 330-foot superseiner. This looked less like a fishing boat than a billionaire's yacht. My contact explained that the skipjack swam in the first 165 feet or so of water, the yellowfin down to 250 feet, then below that, at 500 feet, the threatened bigeye, still shallow enough to be caught in the vast encroaching circle of the purse-seine net, 6,000 feet in circumference and 800 feet deep.

The tuna vessel would be monitoring the location of its FADs by satellite phone. Crew members would be scanning the horizon with binoculars or checking the bird radar. When the satellite buoy found the thermocline, the FADs would be swarming with tuna. Sometimes these FADs might be crude constructions of four sticks and a couple of pieces of wire, and there would be 250 tons of tuna down there, he said. In some circumstances, there could be forty vessels hunting the same fish, and only one would get them. Hence the critical advantage conferred by the best long-range sonar.

In the haul we were looking at on screen, whales were the FAD. Whales, my contact explained, not entirely believably, usually broke the net. It was okay to break the net because of the value of the tuna one still caught. The business was "enormously profitable." This haul alone was 80 tons—worth $235,000—enough to buy a new sonar. But what gave me a thrill of discovery, as I had been looking for it so long, was what was on deck—the bycatch. Gasping their last breaths as the crew pushed the 3-foot skipjack tuna down a chute to be frozen were two 15-foot manta rays.

Jacques Cousteau had a thing about manta rays. The inventor of the aqualung featured them regularly in the credit sequences of his 1960s and 1970s' television series *The Undersea World of Jacques Cousteau* to symbolize the mysteries of the deep. Divers get excited about these black, wraithlike plankton-eaters with their billowing fins. Like barndoor skate, mantas reproduce very slowly, giving birth to just one very large pup every two or three years. Not enough is known about them to say if they are endangered, but the Red List says "populations will rapidly decline unless fisheries are carefully managed."

Manta rays and sharks were a common bycatch, said my purse seining contact. But the biggest problem was the capture of small tuna of all three main species. These were so small that they were of no great commercial value. What sort of proportion of the catch? "They never say what they put over the side." The value of the very latest echosounders, the kind he was selling, was that they were able to measure the size of the fish before you caught them. The only flaw in this argument is that no one says you have to use one.

So we have established that the vessels catching tuna for cans in the Indian Ocean regularly catch at least one vulnerable, large, slow-growing animal, the manta ray, and that they frequently kill bigeye tuna, which is recognized as being on the slide toward extinction. I can now reveal something else, which is that juvenile bigeye tuna—a fish that according to IUCN is as endangered as the Amazon river dolphin, the basking shark, or the North Sea cod—end up being canned along with yellowfin and skipjack. A contact at the IOTC told me that this was "standard commercial practice worldwide." A tuna buyer at Prince's (the canning firm owned by Mitsubishi) openly confirmed this. Perhaps that accounts for the reluctance of some suppliers to say what kind of tuna is in the can.

We are still at the tip of a large sea mount as far as the number of living creatures killed by the purse seine fishery is concerned. I am indebted to a friend for sending me the only observer reports ever published on the bycatch of the purse seine fleet in the Indian Ocean, which came from observers with the Russian purse seine fleet. The paper, by E. V. Romanov, was astonishing. Romanov estimated the catches of tuna in the whole western Indian Ocean at 236,500–313,500 tons between 1990 and 1995—less than it is now. In the process of catching this, purse seiners also caught up to the following amounts: 2,530 tons of pelagic sharks, 1,870 tons of rainbow runners, 1,815 tons of dolphin fish, 1,320 tons of triggerfish, 297 tons of wahoo, 220 tons of billfishes, 143 tons of mobula and manta rays, 88 tons of mackerel skad, 27 tons of barracudas, 176 tons of miscellaneous fish, and an unspecified number of endangered turtles and whales. Quite a high number of sets were on whales. Many recorded were on whale sharks, the world's largest fish, which have been heavily protected since 2002. Although the Russians fished slightly different and in different places, the figures are perhaps the best indication ever compiled of what purse seine fishing for tuna, setting on floating objects, really does to the marine ecosystem. The Russian purse seine fleet is still fishing but is regarded as illegal by the IOTC, which does not allow it to use members' ports.

Then, to his credit, David Ardill of the IOTC wrote to say he had reexamined his rules and decided he would provide information. It was thanks to him that I received, from a kindly source, the most comprehensive and eye-opening answer to the question of what gets killed alongside the skipjack tuna that ends up in your pantry. It was an unpublished paper by two scientists from 1998–99 when the three European organizations of frozen tuna producers declared moratoria on fishing in certain areas close to the African coast because of the numbers of (endangered) juvenile bigeye tuna they had been catching. The authors of the paper note that the moratorium months chosen did not have much effect as they did not actually coincide with the *maximum intensity of fishing in the Somali area.*

The first thing to remark on was that the Spanish fleet was up to its old tricks again and fishing in the waters of one the poorest countries in the world, which currently has no government and therefore is poorly equipped to manage its own waters. The second thing that stood out as the fleet, carrying forty observers, made its way across the Indian Ocean, fishing on free schools of tuna, FADs, and sea mounts, was that some 20 percent of the catches were of the endangered bigeye tuna. The observers also noted the scientific names, in Latin, of the species caught as bycatch. I looked up the common names for these species with growing disbelief. It amounted to almost the entire cast list of *Finding Nemo*. Let's start with the turtles, as these are the most endangered. There was a full house of the oceangoing turtles, loggerhead, green sea turtle, leatherback, hawksbill, and gulf Ridley. These range in their entries from critically endangered to vulnerable in the IUCN Red List. Then there were the whales (remember the Indian Ocean is supposed to be a whale sanctuary): minke, humpback, and one the observer could not identify (a pygmy blue is possible in those waters, but it could have been anything). After that we come to the fish. There was quite a list, topped by the great white shark, which is now officially recognized on the IUCN Red List as vulnerable. Nothing else was on the Red List in a critical category but many, such as the mobulas and mantas, were listed, showing there was scientific concern about their populations but nobody had yet compiled enough information to list them as endangered. What they did say about these species was that for most of them their resilience to overfishing was extremely low. They reproduced slowly, having a handful of young a year. The same was true of the stingray, spotted eagle ray, shortfin mako, great hammerhead, scalloped hammerhead, smooth hammerhead, pelagic sharks, and bigeye thresher.

On the bright side, they didn't catch any dolphins.

How long these bycatch species have before they get into trouble is hard to say, because no one seems to know they are being wiped out, or in what numbers. But there is plenty of reason to be concerned. The concerns about tuna fishing are not just ecological. They are economic, too. Though the tuna fisheries operate out in the middle of the ocean, in the last place on earth where you might think there were limits, there are still no exceptions to Michael Graham's "Great Law of Fishing," that fisheries that are unlimited become unprofitable. Entry to the tuna-catching business is restricted by the tuna management bodies, some more effectively than others. But the right of anyone to buy a vessel and go fishing is still enshrined in international law. Fleets have grown, and most vessels are operating below their economically optimal capacity. The market was the first to deliver a warning about the world's tuna fleet, both purse seine and long line, being far too big in 1999. Catches in that year expanded to 4.4 million tons, and the market collapsed due to oversupply. The catch returned to around 4.07 million tons in 2000 and 2001, but the catch remains unlimited and may rise if there is more demand. Demand goes on rising. Tragedy is inevitable.

A dire warning about the likely future of tuna stocks was delivered by a senior tuna scientist in a paper for the UN Food and Agriculture Organization. Dr. James Joseph pointed out that the human population was likely to grow to 10 billion people by the middle of the twenty-first century, placing even greater demand on the world's natural resources. If there was going to be tuna for them to eat, there was an urgent need to find ways of recording, measuring, and limiting the size of the tuna fleet. "The current legal and political basis ensuring the right of every person to fish on the high seas must be reexamined and brought in line with current reality," he wrote.

Limiting access to common fish resources might best be achieved by giving fishermen property rights over fish in parts of the ocean or making them buy such rights. Currently none of the regional tuna organizations has the legal authority to allocate or sell rights, nor, one might add, to police their seas against vessels that are not signatories to their treaties. What was needed, said Joseph, were "bold new approaches." Time, he emphasized, was limited. Action should be swift.

So coming back to the tuna on your plate, whether caught on a line or in a net, there appear to be at least three major problems—on top of the mercury issue currently preoccupying the U.S. media. Very little action has been taken to restrict the capacity of the tuna fleet, some of the authorities have not even attempted to manage the stock, and hardly anyone is doing anything about the bycatch of endangered species, let alone the unexpected ones actually in the can. Few of the canning companies involved seem to be investing in conservation, and some have an ingrained culture of secrecy about bycatch. At the moment, their policy appears to be: *We can't be blamed for what we don't know is being destroyed.*

Well, it is possible they do know and aren't telling, and if they really don't know, it's time they found out. We consumers will have to consult our own consciences and take what action we can. After this particular journey, I for one will never be able look at a can of tuna in quite the same way again.

13

THE PROBLEM OF EXTINCTION

So long and thanks for all the fish.

—Douglas Adams

Until a meeting at London Zoo in 1994, no one had figured out whether a commercial fish species such as the Atlantic cod faced a higher or lower probability of becoming extinct than, say, the wandering albatross. The realization by many people outside the cozy little world of fisheries "management" of what a mess the world's oceans were in dated from that moment. A lot of terrestial biologists working for the IUCN had read about the Canadian cod crash and felt that they ought to see how fish rated against other endangered animals. Dr. Georgina Mace of the Zoological Society of London remembers: "IUCN had spent four or five years developing listing criteria and somebody said, 'This won't work in the oceans.' We thought, wouldn't it be nice to get the main people together to tell us *why* it wouldn't work." She didn't realize at the time the scale of the conflict she was getting into.

London Zoo had funding for a meeting of thirty scientists

from relevant disciplines in species survival to compare the data for fish populations against IUCN's established criteria of extinction risk. "Two weeks beforehand, the word got round all the fish management people," she told me. The word *tuna* was on the list, so the Japanese said they wanted to be there. Then Spain, Portugal, and Australia said they wanted to send representatives. The fishing nations had realized that a process was going on that was outside their control. Mace remembers: "The Japanese were just too persistent. They wore me down. They came under a whole set of conditions; it was a scientific meeting, they had to bring new data. They didn't protest at the meeting, but the moment they got home they began to attack the thing. [Fishing] is a huge global industry and they were not about to let it fall apart."

A lot of people were unhappy about the idea of listing wide-ranging species with a high reproductive rate, such as cod, as in danger of extinction. Among them were marine biologists who looked at larger, slower-reproducing fish, such as sharks, which had previously been the only focus of concern. Some scientists said that it was theoretically possible for a cod laying 10 million eggs to repopulate an ocean. Mace was resolute: "Any species where the number of adults is going down, must be in trouble. Do you have criteria which pick that up when it's too late or when things are getting worse?" The IUCN criteria say that if a species has declined by more than 20 percent over ten years, however big its population, it is vulnerable to further depletion. Cod and haddock had declined by more than 20 percent over ten years, as had bigeye tuna, so all three went on the Red List, alongside the cheetah, the wandering albatross, and the large-leafed mahogany. Elodie Hudson, who worked with Mace, went to a world fisheries conference in Brisbane. "It got rather hideous," she remembered. "We went through the list over a day, with lots of shouting."

Canada spent a lot of money hosting the IUCN General Assembly in 1996. The Canadian Department of Fisheries and Oceans found out that the 1996 Red List was going to list Atlantic cod, and the IUCN was asked to withdraw its list. Georgina Mace was forced to meet someone from the DFO, who said he was outraged that the whole assessment had been agreed without reference to the scientists who "managed" the cod. The Canadians, for whom everything to do with fish is political, asked how an assessment could possibly be done without calling in all shades of scientific opinion. "We said it was very simple," said Mace. "We just do it." Then the Japanese got up in the first session and protested about tuna. She remembers that there was a lot of bad behavior all around, but some fishing countries were "duplicitous in the extreme." She reflected after it was over: "The population assessments that caused the most controversy were based on UN Food and Agriculture Administration [FAO] data. You can dispute the risk of extinction, not the data."

There was indeed a fair argument to be had over whether the IUCN's criteria exaggerated the extinction risk for commercial cod and haddock, or for other fish stocks of which millions of individuals still survived in the oceans. There is no recorded case of a marine fish species going extinct, as there is of large sea mammals, such as Steller's sea cow. But, then, how would you know the last fish of a species has died out? There is something known as commercial extinction, not a very exact term, meaning that a fish has declined to such a level that it is not worth fishing for. Yet that term is less useful for species such as bluefin tuna, where the price goes up the fewer fish there are in the sea. Fish such as cod are distributed over a large area, so true extinction may not be that likely. Or is it? Canada's cod has proved the existence of the Allee effect—when small populations become less productive—in fish. There are terrestrial species that are like cod in population terms, such as locusts and plants. The North American passenger pigeon was the most plentiful bird species on Earth, yet was gone in five years. The American bison was nearly wiped out by hunting. Elodie Hudson says, "You can't assume that extinction is any less likely in the sea." That hasn't stopped scientists from big fishing countries from stomping around trying to develop endangerment criteria that exclude fisheries.

The first encounter in what promises to be a long war between two sides who each think they know better—fisheries people and wildlife people—was a win for the furry animal brigade. The big bang of the cod crash of 1992, which made few front pages at the time, has rumbled on or more than a decade in the mind of the public, in the pages of *Science* and in disciplines far removed from fishing. People as different as economists and television producers have began to realize that oceans are interesting and that fish are not just scaly things that don't appeal to animal welfare groups but beautiful and fascinating creatures. Political scientists have realized that conserving fish is one of the world's most challenging intellectual problems.

Slowly, a lot of people are asking themselves the big question: if palatable fish are not going to be an infinitely renewable resource, what is everybody going to eat when the world's population reaches 10 billion? For that matter, if the fish are going to run out, isn't that a concrete example of a limit to human growth—as the authors of the Club of Rome report warned in the 1970s? And everyone said they were wrong. Maybe they were right in their Malthusian assumption that human society would need to limit itself or face mass starvation. Or will fish farming and technology make up the gap? Those are things we will discuss, but first let's try to simplify things even further.

It shouldn't be that difficult to run a fishery sustainably. You would think it should be easier than splitting the atom, landing a rocket on the moon, fixing the hole in the ozone layer, managing a currency, or decommissioning the world's nuclear arsenals. But if the failure to do it is anything to go by, it's harder than any of these tasks. How do I know? Because in the course of writing this book I have flown around the planet and surfed the Net looking for a single country where they have done it without making terrible mistakes that crashed some of their stocks, sometimes irrevocably, and I have been mostly unsuccessful. Even in the countries where they are doing okay, there are huge problems, question marks about the future, and failures to think about the survival of *all* living creatures in the sea. So how can we say anyone is yet managing the ocean sustainably?

The problem is relatively simple. People have gotten used to using space technology and weapons of mass destruction in our marine ecosystems, but we haven't developed workable ways of limiting them. We just pretend we have. We haven't even figured out what is down there that we wish to protect, let alone what our grandchildren would resent us for not saving. What makes it hard to manage the oceans is that fishing, the most destructive thing going on in them, is nearly everywhere the responsibility of the most junior politician in the cabinet—the fisheries minister. In some countries he or she is not even in the cabinet. Everywhere in the world the fisheries minister is there just to perform the traditional role of keeping the fishing industry happy. If you've read this far, you would probably agree that this approach isn't working. The question is what to do next.

Here it must all get personal, for otherwise it would get too complicated. I am a journalist by trade and therefore carry in my mental toolkit the opinion that experts may at all times be talking utter nonsense. But you have to start somewhere. It is foolhardy to ignore the wisdom of people who have spent a lifetime thinking about a problem, even if they have made mistakes. They might, conceivably, have learned something. So I will begin by examining the wisdom of the organization that ought to have been fixing the world's fisheries problem but, sadly, has not managed it yet: the UN Food and Agriculture Organization in Rome.

The FAO has been rightly derided for believing bogus information from the Chinese and failing to blow the whistle when the world's fisheries went into decline. But it continues to collect mostly good-quality information and does much good work. It comes up with codes of conduct for responsible fishing and plans for dealing with illegal, unregulated, and unreported fishing, which nation-states do too little about. In its shamelessly pro-development way, the FAO tells the truth, though not necessarily all the truth or all the time.

Think about the FAO's current warning that 75 percent of the world's fisheries are fully exploited, overexploited, or significantly depleted. That is about as close as a pro-fisheries organization is going to get to saying that the world's fish stocks will go to hell in a handbasket unless we do something about it. But consider the FAO's statement that the western central Pacific and the eastern Indian Ocean have the only underexploited fish resources on the planet. Surely what the FAO means is that the fish populations of those oceans are relatively healthy compared with every other ocean. So should a responsible UN body really suggest that more fisheries be developed where there are fewer rules than anywhere else on Earth? As long as you remember that the FAO betrays a verging-on-insane bias in favor of commercial fisheries, you may read on.

The FAO decided to consider whether there were common trends in fisheries that were becoming unsustainable and, if these trends could be found, whether there was anything one could do about them. So it invited some of the world's most eminent fisheries scientists and economists to Bangkok in February 2002. At least the food must have been nice. The centerpiece of the discussion was a paper by Stephen Cunningham and Jean-Jacques Maguire that hit a most un-FAO-like note of alarm. It said that the FAO's current assessment of the world's stocks—that 75 percent were fully exploited, overexploited, or significantly depleted—was likely to be conservative. "Considering the number of [fish] resources on which little or no information is available . . . the situation may be even worse than it appears."

If you wade through the conclusions reached by these fine minds, you will find analysis characteristic of the FAO at its best. First, all the experts present agreed that the factors contributing to unsustainability—that ugly but necessary word—in fishing were similar in almost all jurisdictions. They identified six types of pressures:

Inappropriate incentives. These include subsidies and other economic incentives.
High demand for limited resources. This is reflected in the rising price of fish as a luxury, healthy, fashionable food.

Poverty and lack of alternatives. This is the hardest problem of all, for if, like penniless fishermen in the South Java Sea, you have nothing and no job, you will carry on fishing long after anyone would have paid you to, so that you can eat.

Complexity and inadequate knowledge. By this the experts meant, I think, both the lack of adequate scientific data on complex fisheries, such as the West African upwellings, and human legal systems that are too complicated to be understood properly.

Lack of governance. This is a good old UN euphemism for corruption, fiddling, incompetence, and failure to enforce the rules.

Interactions of the fishery sector with other sectors and the environment. "Interaction" is a characteristic FAO euphemism for messing things up, as fishing surely has.

The gurus of huntering and gathering also came up with a list of new ideas intended to address the world's problems. A few of these are platitudinous, but not many. I'll give their headings, then translate, as before.

Rights. The FAO's worthies were of Garrett Hardin's view that we should fence the commons by granting property rights. They called for the allocation of "secure rights to resource users (individually or collectively)" for use of a portion of the catch or a portion of the space taken up by the fishery. This is a vast change, the implications of which are discussed in Chapter 14.

Transparent, participatory management. This doesn't happen in many places right now, though some countries are already well on their way. Iceland, for example, gets its fishermen to do an annual scientific trawl survey, which means they are more disposed to believe its conclusions.

Support to science, planning, and enforcement. A bit platitudinous, this.

Benefit distribution. Now this is an interesting one. It means that fishermen should pay rent for using the fishery to the people who own it, that is, the public, possibly on top of the income tax they pay. This works only if the fishery is profitable, which only those run on economically efficient, though not socially inclusive, lines are. This is an important suggestion that would compensate to some extent for the downside of handing out rights to the fishery, which over time would concentrate quota in just a few hands.

Integrated policy. I guess this means that each nation should set explicit objectives that address all aspects of sustainability, such as the conservation of marine wildlife as well as fish. This would mean the creation of some marine reserves. So why not say so?

Precautionary approach. This, I suspect, is UN diplomatic code for saying that none of the world's fisheries institutions should be managing stocks for maximum sustainable yield anymore, as many tuna bodies are still doing. They and many other fisheries should be setting total allowable catches, according to FAO guidance, so that fish stocks would recover even if the worst climatic or population fluctuations previously recorded were to hit the stocks.

Capacity building and public awareness raising. Again, necessary but platitudinous.

Market incentives. By this they meant they agreed with

ecolabeling schemes for the consumer that identify sustainable fish against unsustainable fish. Not a bad idea at all.

The assembled gurus also had some good ideas about governance. They made a strong plea for the creation of institutions to manage the high seas where these did not already exist. Where they did exist, they said these institutions should be strengthened considerably to play their necessary role. Each should receive, *as a minimum,* clear legal authority to manage and conserve fish stocks, and both the political mandate and the resources to ensure that fishermen complied with conservation policies and regulations. They meant that the day-to-day requirements of fishery managers should not be politically negotiable, as they are in so many countries. Thinking about their recommendations, I can name few countries in the rich North, let alone the rich South, that have institutions of this kind today. So this is quite a recommendation.

As a way of speedily immersing oneself in the latest ideas about saving the world's fish, this summary is hard to beat. But I must add two observations from my own journey around the world: the problems are worse than many people imagine and the solutions more controversial. It is one thing, for example, to recommend the use of rights to control fisheries, another to persuade people to accept them over time and make them work— as we will find out in the next chapter.

You miss the gossip, the sense of what the key points were, by merely reading conference conclusions. So I spoke to Jake Rice, one of the participants. Rice is Canadian and lives within a culture that considers fishing a cultural (i.e. subsidized) activity as well as an economic one. He goes misty-eyed about the guys trying to make a living in all those outports in Newfoundland—though they are subsidized up to the hilt.

From something Rice said, I suspected there was a conclusion they all came to, over the Thai food and beers, that was too difficult for the FAO to write up. Jake said as much himself. He said you wouldn't find this bald and heartless sentence in the conclusions because the FAO, being a UN body, had to have "some sense of hope in its texts." This, however, is what they really concluded in Bangkok: "You have to be willing to write off one of three dimensions—ecological, economic, or social—to solve the problem of sustainable fisheries."

Let's examine that proposition, because this is as close as we are going to get in fisheries to a solution. There are three options:

A. You can have a fishery that is healthy ecologically and economically, but you have to forget about supporting fishing as a cultural (i.e., subsidized) activity all the way around the Newfoundland, Scottish, or Iberian coast.
B. You can have a fishery that works economically and socially but not ecologically (but presumably not for very long because the resource will be gone in a few years once you have mined it).
C. You can have a fishery that works ecologically and socially but not economically (conceivably by making its money out of something else, such as tourism or nature conservation grants, like paid graziers on a nature reserve).

Only A and C work because B is probably not sustainable in the long term. My view is that A may well have a future in offshore waters, while C may have more of a future in inshore ones. I think this tacit understanding among the FAO guys is actually quite profound. It is an answer to the meaning of life in a world of constantly expanding technology.

It may be profound, but it is not wholly original. Jake's workshop aphorism reminded me of something John Pope, then a senior scientist at Lowestoft Laboratory and subsequently an adviser to the Canadians on their cod stocks, told me in 1997: "You can't maximize the economic and the social and the ecological all the same time." In other words, something's got to go.

So what's got to go, you may ask, if we are to have fish forever? I think the answer, almost certainly, will be different in different places. But in most industrial fisheries more than a day's journey from the shore, I'm beginning to think I know what's got to go, because it's gone everywhere else in raw materials industries the world over. And that's people.

Why should fishing be any different from mining, shipbuilding, or farming, especially if it's destroying the world that belongs to the rest of us? What is special about outports that means they should survive? The bottom line is that there are just too many fishermen, and fishing technology gets better every year.

Jake Rice believes in the cultural side of fishing. He thinks there is something we want to hold on to in coastal communities. He thinks there is something about those outports that needs to be preserved *as it is*. He may be right. But I am not so sure I want to preserve the Scottish, Iberian, Newfoundland, or former New England kleptocracies one little bit. So I will let Rice, who is a more compassionate soul, make the argument himself, for he made it beautifully, and far better than I could, in an e-mail he sent me:

I grew up in an area of dairy farms, with small towns scattered every few miles apart—each providing the feed mill, milk plant and general store for the neighboring farms. Now agribusiness owns more than 90 percent of the farms, and the little towns are either empty or gentrified by people who commute an hour each way to the city to work and shop. A rich, interesting culture was lost, and no one blinked. Other examples abound. It is a genuine mystery why coastal fishing communities have such a uniquely sacred identity, but the pattern repeats in Canada, the USA, the UK, Denmark, Iberia and the rest of Europe at least. I really don't know why—maybe better musicians and songwriters.

Now we're getting somewhere, I thought. My view, as the son of a man who was both a farmer and a miller, whose family's businesses are no more, is that we live in a different world. There is plenty of employment, if you adapt, without having to hoe cabbages by hand, which I did in my youth. You can't go back. Rice, who contributed to the crash of Newfoundland cod but acquired some wisdom in the process, is a wise but incorrigible sentimentalist. He loves fishermen, just as the world used to. But can we afford to be sentimental about fishermen if the price of having too many of them is the destruction of fish and the wonders of the oceans, which we are only just beginning to understand?

We need a new approach. People make a virtue of necessity only if you stop subsidizing them to do the wrong thing. In Lowestoft, where there is no fleet and very few fish any more, they produced the Darkness, a retro glam-rock band that briefly took the world by storm, and tourism is picking up the slack. We used to think miners should be subsidized. We know farmers *are* subsidized but realize it is a disgraceful subsidy. Very soon we will think the same about subsidizing fishing.

But that's enough theory. I really think it's time for a reality check, to see if these ideas work somewhere out there where they are trying to save the cod.

14

DEATH OF THE COWBOY

Reykjavik, late September. Flecks of the first snow of autumn drifted out of a leaden sky above the Hvalfjordur, the whale fjord. Here on the capital's waterfront, one of Iceland's big freezer trawlers, the *Therney*, lay offloading her frozen, filleted cargo of redfish and cod in cardboard boxes labeled in Icelandic, English, Spanish, and Japanese. Fishing forms a significant part of the economy in Iceland, and, as a result, the stock market takes the annual fish stock assessments very seriously, dipping if the cod assessment is down. Kristján Loftsson, a part owner of the *Therney*, who brought me to the dock in his huge Jeep, a characteristic vehicle for a well-to-do Icelander, pointed to the front page of *Fiskifréttir*, Iceland's fishing newspaper. It showed an Icelandic trawler chugging back to port, its hold so full of blue whiting that the boat sat low in the water and looked like a submarine. Loftsson shook his head. The picture annoyed him because he saw it as a stain on Iceland's reputation. Icelanders like to think they know as much as anyone about conserving fish.

They are almost certainly right. Icelandic waters are among the few places in the world where cod stocks are both relatively plentiful and, arguably, on the increase, though the upturn is in its infancy. After several years of setting quotas low, and enduring criticism for doing so during a general election, the government had just agreed to increase the quota by 33,000 tons, in line with scientists' assessment of what stocks could stand and their belief that big year-classes were beginning to come through. This increase in allowable catches was equivalent to two-thirds of the entire spawning stock of cod in the North Sea. Here in Iceland it was a kind of banker's Christmas bonus, bringing the total allowable cod catch up to 229,900 tons, and making those people who own cod quotas even richer than before. It amounted to a windfall of $93.3 million for the Icelandic economy, a tidy sum among a population of only 300,000 people. The injection of cash pushed prices higher than they were already, and moved Icelanders further up the list of the richest people in the world.

Iceland's ability to manage its fish stocks starts with its isolation. The distinctiveness of the place hits you when you get in the shower and smell the sulfur in the water from the hot springs. Iceland's ability to determine unilaterally what goes on for 200 miles around its coast does not necessarily make for good management. The years from the end of the last cod war in 1976—when this country without a navy expelled the British and German trawlers that had overfished its stocks of cod—to 1984 are now just a bad memory. It was after the last cod war that overfishing by Iceland's own trawlers, some bought from Britain, replaced overfishing by foreigners, as happened in Newfoundland after the declaration of their 200-mile limit. Iceland took a stronger grip of the situation than Canada, proving perhaps that countries where fish represent a sizeable proportion of their income take more focused action than countries with more numerous ways of making a living and only a social interest in fishing. "In Iceland we know that if we mismanage fisheries, we are out," said Hjörtur Gíslason, who writes about the fishing industry in *Morgunbladid*, the leading Icelandic newspaper, and was the first European journalist into Newfoundland when cod fishing was banned in 1992.

It hasn't been easy. Iceland's Marine Research Institute publishes a Blue Book every year showing the state of every stock in Iceland's territory, from whales to whelks, including the real revenue earners, such as cod and haddock. The Blue Book is a great exercise in transparency and public information. If you look in it, you will find that cod, Iceland's most important stock, was formerly much more plentiful than it is now. In the mid-1950s the fishable stock was estimated to be over 2.2 million tons and the spawning stock over 1.1 million tons. There were some nasty shocks as a result of overfishing from 1980 to 1983 and in the late 1980s. Consequently, in an effort to rebuild the stock, the annual cod quota was reduced drastically from over 330,000 tons to 220,000 tons. Then, around the late 1990s, scientists drastically overestimated the spawning stock, a mistake they realized and admitted in time. Iceland cut back its cod quota even further, to 165,000 tons, which caused some complaints about bad science, but the fishing industry tightened its belt and put up with it in anticipation of better times ahead. For three years, one during a general election in which the ruling coalition was re-elected, the resolve held firm at the ministry and the institute. At the time of my visit cod numbers appeared to be rising again.

There are striking contrasts with other parts of Europe. In Iceland, where the fishermen are mostly rather rich, they accept the science, like major public companies dealing with pharmaceuticals would, instead of perpetually grumbling, as their poorer European brethren do, that there are more fish in the sea than the scientists see in their trawl surveys. Does this tell us something? Who honestly challenges the science—rather than the interpretation of the science—in any other discipline and is taken seriously? Would you have an operation by a surgeon who didn't take the science seriously? Would you buy a car built by someone who didn't? Jóhann Sigurjónsson, director general of Iceland's Marine Research Institute, told me: "Iceland is perhaps the only European economy that is still based mainly on hunting. We are heavily dependant on the resource. So Icelanders have accepted in principle that science-based management is the only way to go. We may argue about it, but at the end of the day there is a consensus. We have general support for our recommendations."

One reason for Icelandic fishermen respecting the scientific process of counting fish is that they themselves are involved in surveying the stocks. For a few days a year a handful of trawlers is supplied with nets of the kind they used in the mid-1980s, which are mothballed somewhere in Reykjavik, and these are used in the annual trawl survey, alongside results from grids trawled by the institute's two impressively equipped research vessels. Some people complain all the same that changes in the climate mean that the fish are no longer where they were in the mid-1980s. They may well have a point. Some blamed the hot summer for the scallop population's collapse—or was the explanation too much scallop dredging? Monkfish, a species once caught only on the warmer south coast, is now caught all around the island. Juvenile haddock, saithe, plaice, mackerel, and cod have all thrived since 1995, when the sea became warmer, but there are also fears that climate change may be interfering with the breeding success of cod in Icelandic waters, and may be the reason why the population is not more obviously increasing. There are many unknowns, even in one of the best-studied fisheries.

You do wonder whether Sigurjónsson and Iceland's fisheries minister at the time, Árni Mathiesen, should have left the quotas lower if their stated aim to let the stock recover and grow to former levels was more than rhetoric. Crucial to this growth was to build up the numbers of all-important older spawning cod— the "old ladies," as Sigurjónsson calls them. He was surprisingly downbeat when I talked to him, and I suspected a political accommodation may have been forced on him. He said: "The stock is healthy. Some people would be very happy. We are not happy unless we are achieving 100 percent of what we want. We are advising now not on a crisis but on the best level we want the fishing to hit." I've noticed that it is at precisely this point that conservation, even in the best fisheries in the world, runs out of steam, because there are so few parallels for holding on tight and restoring the stock to its former abundance.

As Odd Nakken, a wise Norwegian fisheries scientist and former director of the marine research institute in Bergen, put it to me a decade ago: "The history of conservation that hurts is only twenty years old."

Sigurjónsson will tell you that fishermen are still catching the cod too young. The majority of those caught are six years old, and cod only begin to spawn at five around Iceland. He would rather be catching the country's quota from cod age seven, eight, or even nine years old. A nine-year-old cod can weigh 11–16 pounds, so the number killed would be fewer for the same weight of fish. Nature equipped the cod to live for over twenty years and to weigh more than 90 pounds in order to spawn longer. The old ladies will lay their eggs for a longer period—

up to six weeks. These eggs, which are larger, healthier, and stronger than those laid by smaller fish, stand a better chance of hitting the optimal water temperatures for survival. For the infant cod to thrive there must be plenty of food. Sigurjónsson has plans for closed areas to protect some of these old ladies. He said the cod recovery beginning then—and that two years later we are still seeing—has been achieved by sacrificing the shrimp stocks, paradoxically worth more than the cod itself until there is a major cod recovery.

I am still surprised to see just how fine the line is between responsible fishing and overfishing, even in a system where politicians and scientists have the mandate to make tough decisions. And how hard these decisions are, given growing climatic variability and a principal stock, cod, that is perhaps a quarter of its biomass fifty years ago. The recovery of cod, haddock, saithe, and plaice is more tentative, the variability just as great and sustainability less ensured, even in one of the best-run fisheries in the world.

Worries aside, the Icelandic system has more precautions, more meaningful penalties, and more flexibility to close areas where juveniles are caught than most others. This extra caution is written into government policy. Since 1995, the total allowable catch of cod has been limited by the government's harvesting strategy to no more than 25 percent of each year's fishable stock. Compare this with nearly 100 percent of spawning adults (and quite a few juveniles) being killed in the North Sea and Georges Bank in the 1990s and you are not surprised that there is barely anything left in those waters. Another admirable precautionary measure built into the Icelandic system is that an annual increase in total allowable catch of more than 33,000 tons is not permitted, just in case the scientists have got it wrong. The

Marine Research Institute also has powers to close fishing areas at a few hours' notice if it believes the fish being caught are too small.

The Icelandic system has a lot of other things going for it, too. The basis of the tremendous economic success of Iceland's fishing industry since 1984 was the introduction at that time of a revolutionary, rights-based system of quotas, which give fishermen a long-term investment in the health of fish stocks—just what the FAO worthies now recommend (more than twenty years later) as the best way of running a fishery. Iceland's system of individual transferable quotas (ITQs) has been as successful as it has been controversial. (Don't worry too much about the name, of which there are variants around the world—"individual fishing quotas" is used in the United States, for example. It just means you own that proportion of the quota for good unless you sell it.) As with any new system, there have been winners and losers. The problem that some people perceive in this one is that it has tended to concentrate the quota in the hands of large companies. People who have played the system skillfully have made large sums of money out of the rise in the value of the quota. A handful left the country and now trade quota on their mobile phones from Florida or the tax haven of Guernsey while doing no fishing at all. Meanwhile, coastal communities that were slow to buy enough quota for their needs, or were already in debt, ended up having to sell their trawler quota. The system in place remains extremely popular with those who operate the big freezer trawlers more than 12 miles offshore. They will tell you that it has, quite simply, changed the way that fishermen think. In the inshore waters there was initially a different regime based, controversially, on effort controls—limits on the time that can be spent at sea. In

2005 this changed, and now all commercial fishing is controlled by ITQs.

Iceland's tradable quotas are not absolute property rights in perpetuity, as were allocated in New Zealand. They amount to a right to take a certain proportion of the stock, depending on scientific advice, for the foreseeable future. This, fishermen say, has brought an end to the race to catch fish before somebody else catches it. Fishermen, therefore, are more likely to leave fish in the sea if they think prices are poor at the moment, for they are sure it will be to their advantage in the long run. As Árni Mathiesen told me at the time of my visit: "You have somebody who thinks of the quota as their property, who thinks it has to be as good tomorrow as it was yesterday. That is what matters in the long run."

Fridrik Arngrímsson, chief executive of the Federation of Icelandic Fishing Vessel Owners, told me: "Since ITQs were introduced, the thinking has been completely different. We don't think of volume so much anymore, we think of value. Once a skipper caught 80 tons and showed it to me proudly. What he did not say was that he was getting nothing for the cod because it was all twisted and bruised." Fishermen who fish for maximal economic efficiency fish less, burn less fuel, and drag up the bottom less as part of the bargain.

Acquiring the right to a certain amount of fish, come good weather or bad, has changed not only the way fishermen think but the way they fish. Rúnar Stefánsson, the manager of the Grandi fleet, which owns the *Therney*, told me that fishing was no longer a rush to catch the most fish, but a matter of catching high-quality fish and selling them for the best prices.

All sixteen of Iceland's major commercial stocks, from plaice to capelin, are now managed under ITQs. People in Europe who say the Icelandic system wouldn't work in a mixed fishery such as the North Sea or along the coast of West Africa don't understand what they are talking about. The rules are strictly enforced by the directorate of fisheries, a government body, and the coast guard. Indeed, one of the major advantages of the ITQ system is that there is an incentive for fishermen to watch their neighbors to ensure that others are complying with the rules, for they will lose out if the fish stock declines and the value of the quota falls. There are "black" landings, but these are thought to be far less than they are in the EU or elsewhere, perhaps under 3 percent of the cod catch.

Stringent rules and enforcement are crucial to the system. The rules say that everything caught must be landed (no throwing away juvenile fish below the size limit). Discarding is banned. Regulations on mesh size are supposed to mean that juveniles are not caught. All the same, Iceland's waters are a mixed fishery, so fishermen are not always going to catch what they set their nets for. The system therefore allows some flexibility. If you do not have quota for what you catch, you may buy it to make your catch legal. Or you may transfer some of your cod quota to catch, for example, monkfish. You may not use monkfish quota, however, to catch more sought-after cod. If nobody at all wants to sell you quota for your over-quota cod, you are allowed to land your fish, and 80 percent of the value of the over-quota catch will go to the Marine Research Institute. The system is backed up with tougher penalties than in other regimes. Skippers who break the rules run a real risk of not being allowed to fish, with serious economic consequences. Licenses can be withdrawn temporarily or permanently. Skippers convicted of serious offenses go to jail.

Supporters of the ITQ system are evangelical about the way

it has improved the psychology of fishing and transformed fisheries investment. "The whole world is heading this way," says Ragnar Árnason, economics professor at the University of Iceland in Reykjavik. "If people have property rights, they look after them." What appear to be the most stable and sustainable fisheries in the world—hoki in New Zealand, rock lobster in Australia, and black sablefish, halibut, and (up to a point) pollock in Alaska—are now run on this basis.

Over dinner in Reykjavik's Hotel Holt with Ragnar Árnason and my friend Orri Vigfússon, a great campaigner for the Atlantic salmon, I asked whether we can go over it again. Árnason said the people who own the rights will defend and protect the system. But is a right to catch a fish the same thing as owning a piece of land? Fishermen in Iceland still pay mortgages and are governed by what economists call the discount rate, that is, it is better to have money in the bank tomorrow than in a year's time. Although I was increasingly convinced by what Árnason says—that ownership is the only system that gives a value to the fish left in the sea—I decided to play devil's advocate. I asked whether there are not just as many successful examples of running a fishery around the world (rare though these are, viewed overall) that are based not on rights but on the principle of a managed common? Vigfússon indulged me while I came up with three examples. I decided to rate them, and Iceland's system, out of ten.

THE NORTH SEA HERRING

The history of the North Sea herring in the 1960s and 1970s is reminiscent of the gold rush on blue whiting today. The traditional way the British and Dutch fished for herring, with a drift net, came to an end soon after the Second World War. The purse seine and the bottom trawl, targeted at spawning concentrations of herring, were the instruments of destruction in an international free-for-all. Norwegian, Icelandic, Bulgarian, Polish, and East German vessels raked the sea beyond the 6-mile limit. Britain joined the Common Market in 1973, placing the whole North Sea under the management of the Common Market plus Norway for the first time, but it was still impossible to reach agreement among all the people who fished for herring.

The stock collapsed in 1977 and the fishery was closed. The next year the much smaller herring fishery west of Scotland closed, too. The fishery reopened in 1981 after a partial recovery. For the first time it was governed by a system of total allowable catches. Unfortunately, the closure from 1977 to 1981 virtually destroyed the market for herring in Britain—people had gotten used to eating other fish—but not in eastern Europe. So vast Russian factory ships, known as "klondikers," with fleets of catching vessels, fishing on licenses bought from the European Community, still fished the herring, as did the Scottish, Danish, and Norwegian pelagic fleets. In 1996, the herring was again under pressure. A 50 percent cut in quota was agreed and enforced because the memory of the crash was still alive in people's memories. It seems to have worked. The pelagic sector is one of the EU's few successes. Its fleet comprises a few, highly equipped boats and appears relatively under control—despite, or even because of, some recent spectacular fines for cheating. Stocks are healthy, though not the great bounty they once were.

Grade out of ten: 5, just for the herring, not the whole EU Common Fisheries Policy/kleptocracy.

BARENTS SEA COD

The cod stocks shared by Norway and Russia in the Barents Sea are the largest left in the world. In the Lofoten Islands, stepping-stones that stretch out into the Atlantic just inside the Arctic Circle, you see racks of dried cod or "stockfish" drying in the breeze in the same way they did when the Vikings built their longhouses over a thousand years ago. The Barents Sea's huge resource of cod sustained the whole of Europe through trade with the medieval Hanseatic League, but it is now somewhat depleted. The total quota agreed by Norway and Russia in 2003 was about 484,000 tons, but one should remember that the Barents Sea used to produce catches of 990,000 tons a year in the 1970s. Lately stocks are down, but not as low as they were in the mid-1980s, when catastrophe was only narrowly averted. In 1986 and 1987, thanks to the overfishing of capelin for fish meal for Norway's salmon farms, the cod suffered a food crisis and cannibalized their young. Contrary to scientific predictions, the cod stock did not grow. In a decade of bullish estimates it declined rapidly. By 1989 the cod was in bad shape, and scientists, led by Odd Nakken, recommended a drastic cut in the quotas. There was resistance from fishermen, but a determined fishing minister, Oddrun Pettersen, from Finmark, the most fish-dependent county in Norway, banned fishing for capelin and slashed the cod quota to 220,000 tons, shared between Norway and Russia. For the first time Norway threw subsidies at fishermen not to fish. It paid the mortgages on the fishermen's vessels.

By 1996, when I visited the Lofoten Islands, there was plenty of cod again. But Odd Nakken was unhappy when I met him for a drink in Bergen. Norway operates a system of total allowable catch and individually nontransferable quotas. Norway and

Russia's ministers had agreed a total allowable catch of 935,000 tons of cod for that year—about 110,000 tons too high, observed Nakken, puffing his pipe. So it proved. Norway's subsidies had sent the fleet into temporary retirement, but it simply got going again. Paying fishermen not to fish without reducing the size of the overlarge fleet merely postponed rather than cured the problem. The cod stock has tumbled since then to a worrying extent. I am told that Norwegian fishermen suspect that the Russians are cheating on their total allowable catches and quotas, so they are avoiding their own fairly stringent rules. Norway is also exporting its subsidized fishing capacity by catching blue whiting in the current free-for-all. Its share of the carnage in 2004 was about 770,000 tons. Many voices in Norway say it would be better off with ITQs. The government does not want them because it says this would be inconsistent with its policy of supporting coastal communities in the north.

Grade out of ten: 9 at the beginning of the decade, but subtract 6 for going wrong since; result 3.

THE FAROE ISLANDS COD

The Faroese cod stock was healthier than some—friends told me of disbelieving looks on the faces of Scottish fishermen when boxes of fat Faroese cod were landed at the Scottish port of Scrabster in the early 2000s beside the puny specimens from the North Sea. Like Iceland's, the cod stock appeared to be increasing for climatic reasons. Faroese fishermen traditionally catch cod by long lining or "jigging"—a technique that involves jerking a silvery lure up and down on the end of a line—rather than trawling. Jigging is now done electronically by several machines attached to the side of a fast inshore boat, making it a light, efficient killing technique. Faroese cod produced catches of around 33,000–44,000 tons a year, reasonably reliably, for years—except in the disastrous years of 1993–8, when catches plummeted to around 6,600 tons for reasons that have never been satisfactorily explained. They then bounced straight back up again. Until the early 1990s, 20 percent of the entire government budget was spent on fishing subsidies. Since then, these have been stripped away.

The Faroe Islands operate a system of effort control, under which each vessel is allocated a number of days at sea. Like Iceland, it also uses closed areas to protect spawning aggregations of cod. The quota is divided up and a committee works out how many days each vessel can fish. Interestingly, the Faroese had a system of ITQs imposed upon them by Denmark after their cod crash. ITQs did not have the backing of the fishermen, so they did not work. After a rethink, controls on effort—time spent fishing—were the industry's choice, and now these seem to work, but economists such as Ragnar Árnason justifiably say the system is less economically efficient. A fishing vessel could find itself sitting in port for half the year. The system restricts a certain size and gear type to particular areas around the Faroes. All fish caught during the trip can be landed and legally sold. There is the usual problem that there is always some aspect of fishing that effort controls do not constrain—whether it is the sonar, the kind of gear, or the power of the vessel. And it is human nature for fishermen to increase these. Sure enough, in 2005, ICES said that fishing effort has actually doubled since the effort control regime began. This was despite the Faroese system of theoretically compensating for technological creep— the annual improvement of gear, boats, and sonar. It lowers the number of fishing days by 2 percent a year.

Something obviously wrong with the Faroese system is the amount of cod it awards the fishermen—33 percent a year of the total biomass of the cod, which scientists from ICES say is too high. The Faroese parliament prefers the advice of its own scientists. ICES says that the Faroes ought to build a more precautionary approach into its stock assessments and make its records more accurate, and that the current rate of fishing is probably not sustainable. Stocks appear to be on their way down again, but given the state of the records, it's hard to tell by exactly how much.

Grade out of ten: 3.

THE ATLANTIC SALMON

As well as these three examples of how fisheries have been managed with some success in a common sea, I can't resist adding the example of Orri Vigfússon's buyout of the Greenland, Faroese, northeast England, and now some Irish drift nets for Atlantic salmon. This example cuts across categories—common sea fisheries and rights-based fisheries, recreational and commercial forms of fishing—but solutions in conservation, I have noticed, invariably come from lateral thinking. The decline of the Atlantic salmon in the 1980s set anglers thinking about the overwhelming number of threats to migratory fish: habitat degradation in rivers, poaching, estuary nets, and, most intractable of all, high seas netting on the salmon's annual migration to feed off Greenland.

Orri, who owns a vodka factory in Iceland, decided to do something about the difficulties the international body responsible for the salmon, the Edinburgh-based North Atlantic Salmon Conservation Organization, was having negotiating downward the numbers of salmon caught in their feeding grounds at sea around Greenland and the Faroes. He single-handedly set up a body called the North Atlantic Salmon Fund and called on owners of river fishing in Europe and North America to protect the value of their properties by buying out commercial fishing capacity. The Inuit fishermen of Greenland were not making much money out of drift netting for salmon, which was stopping several hundred tons of salmon from returning to European and North American rivers, where they would have made considerable revenue for the owners of sport fishing rights. The Greenland fishermen were happy to be handed money and grants, through their fishing association, to reorganize themselves to tap more profitable fisheries.

A deal with the differently organized Faroese fishermen followed, and a deal with fifty-two out of the sixty-eight licensees of drift nets off the northeast coast of England (worth $5.4 million to the fishermen) followed that in 2003. Orri's river in Iceland has seen salmon numbers soar following the Faroese net buyout. The River Tweed, on the border of Scotland and England, happened to have its best year since the 1960s after the northeast drift nets were bought out and has done well ever since. The key to all these deals seems to have been that Orri and his supporters negotiated with the private interests that exist even in a common. In some cases, like the English North East fishery, the law actually had to be changed to give the commercial fishermen secure rights before these could be bought out.

Grade out of ten: 9.

HOW DOES ICELAND COMPARE?

Before Ragnar, Orri, and I attempted to compare how Iceland rates against the Faroes and Norway, we discussed the downside of the Icelandic system, which I had been talking about with opposition politicians. Because of the way quota allocation is decided, the acceptance of this system within Icelandic politics is more fragile than outsiders imagine. Kolbrún Halldórsdóttir, an MP of the small Left-Green party, put the problem succinctly:

> When the quotas were given out, they were a gift. Then they got a value. Big companies bought the quotas from small companies, and the small villages had no fish coming into their harbors. That became a reason for people moving away from the villages and into small towns. It angers us to see people who are filthy rich and living in Guernsey putting their money into shopping malls in Reykjavik instead of into the fishing industry. Sustainable development is not about collecting money into one big pile. You have to reduce the ability of the company to access it in big piles. You have to recognize that the plan does not work and recognize the imbalance it has created in our society.

The Left-Green party has come up with a thoughtful way of restructuring the system of ITQs over twenty years so that the quotas are tied to communities rather than companies. This has been received as ingenious by all shades of opinion. But, crucially, it is still based on rights.

Other opposition parties look to the Faroes for solutions. Magnús Thór Hafsteinsson, an MP for the Liberal party, told me: "There is a human factor in all this. The small communities that depend on fish are losing the battle." Gudjón Kristjánsson, a Liberal MP and former skipper, said: "The winners are always winning and the small ones always losing. The fisherman can sell his quota, but who will buy it? The big company."

Hafsteinsson points out that the system of tradable quotas was meant to lead to a contraction in the fleet, whereas the decrease in the number of large trawlers has been disappointing. This is a good point. In fact, the number of licensed vessels has declined by almost half since 1984 and the fleet tonnage has also declined, but the decline in the fleet has been much less than you would expect from the reduction of fishing effort within Icelandic waters. The reason is simple: many of Iceland's fleet of big, superefficient 230-foot trawlers now fish both in Icelandic waters and for redfish, blue whiting, and other species in effectively unrestricted fisheries beyond the 200-mile limit. Iceland has turned from a coastal waters fishing nation to a buccaneering force upon the high seas, not yet like Spain, but a different kind of fishing nation from the one it was. The Liberals would like to move to effort control, with the effort divided increasingly between the smaller vessels.

This remains a popular message, and one that was actually happening by default in Iceland, but it threatens the entire ITQ system. Back in 1983, when transferable quotas were handed out, the politicians of the time bought peace with the owners of hundreds of small inshore vessels around the coast by allowing them free fishing. The inshore vessels originally took only 3 percent of the catch, but, strangely enough, that percentage grew and grew. A series of rules restricting the size of inshore vessels proved counterproductive, as prompted a technological revolution. Fishermen sold larger boats and bought faster, smaller ones equipped with the very latest in powerful engines, electronic cod jigs, and other gizmos. By 2004, the coastal boats were taking more than 25 percent of the quota. They always caught more than they were supposed to, but 25 percent was still not enough for the genial giant Arthur Bogason, who represents the inshore fishermen.

Charming and "green" himself, Arthur wanted more for his people, and over a delicious lunch at his expense in one of Reykjavik's finest hotels he very nearly convinced me that he should be given it. Unfortunately, it is the job of writers to bite the hand that feeds them, and I must report that if there is one part of the Icelandic system that was out of control, it is Arthur's. The inshore fishermen had become greedy, and they were undermining the sustainability of the system. Although the inshore boat is capable of producing top-quality fresh fish, which is then whisked to the shore within the day, the system of effort controls meant that there was an inbuilt temptation to catch as much cod as you could and rush it home without icing it properly. Interestingly, the year after I was there, Arthur Bogason's association of inshore fishermen decided that their interests would be better served within an ITQ system. With their tacit agreement, a law to that effect was passed by the Icelandic parliament in 2004. Ragnar Árnason reports that there has already been some consolidation of ownership among the holders of quota.

Iceland, when I was there in 2003, was a crucible of competing theories for controlling fishing. Which system was the best? I think on the evidence you have to conclude that the system of tradable rights has the edge over effort controls because it gives a value to fish that are left in the sea. If rights work, they give fishermen a financial interest in conservation. They also give the fishermen who hold quotas an interest in reporting other fishermen who break the rules, and in fishing in a different, more careful way. In political terms, having an electorate of owners rather than supplicants takes part of the responsibility for achieving conservation away from politicians and vests it in companies, which is no bad thing, as politicians have failed everywhere in their duty to conserve fish. Rights-based systems take the responsibility for decommissioning away from treasuries—which never want to fund this anyway—and give it to the market. It is easy to think of circumstances in which a rights-based system might not work, such as if fishermen resented the way it was imposed, as they did in the Faroes. Rights are not a panacea for all the ills in fisheries. On the other hand, a private company in New Zealand went to the fisheries minister and asked for a *reduction* in hoki quota. Where else would that happen but in a rights-based system?

Effort controls, by contrast, give an incentive to fish as hard as possible in the time allowed. And they encourage technological creep. In Ragnar Árnason's words, "Effort reduction does not work because fishing effort has so many variables. You can always find the dimension of effort that is not constrained and not constrainable." The ambitious, high-tech inshore fleet—which surfed on a wave of nostalgia in Iceland for a time when fishing was done by small entrepreneurs, not major businesses—was a problem Iceland had to address before it undermined the whole system. I leave it to the journalist Hjörtur Gíslason to nail the problem: "The trouble is there is not enough fish for everybody. It does not matter which system you have. That is limited."

Grade out of ten: 8, because I still have anxieties about the way the allocation of quota was done, the lack of marine protected areas, and the tentativeness of the cod recovery.

So much for Iceland, but what about the rest of the world? Do ITQs have to be as socially divisive as they have been in Iceland? I asked Ragnar Árnason. It's not great advertising for a system of property rights at sea if one sees a few fat cats, in Hjörtur Gíslason's phrase, "frying their bellies in Florida" in twenty years' time. Árnason remains relaxed about the inequalities of wealth that have come from the share-out of quota in 1984: "It is a fundamental point of human nature that you can't create new wealth without a quarrel over how to share in those benefits. The ITQ system increases the economy overall, so many people will benefit. It is the most efficient way biologically and economically of using the resource. It does not solve poverty, but then what industry does?" Once the quotas are profitable, they can be taxed, like any other area of the economy, and the revenue used for programs to help the poor and dispossessed. Possibly reflecting the same sentiments, Iceland introduced "rents," that is, a special tax on ITQ holders, in 2004.

That didn't quite tackle the problem I foresaw, so I asked the question a different way. Could he really live with the kind of inequalities that the ITQ system seems to engender in someplace such as Senegal, where the alternative for nonfishers is starvation? Árnason replied: "You can solve all equity problems at the outset. You can choose who to give the quota to, who is eligible. The ITQ system is equity neutral. You can achieve any desired distribution of the benefits through the initial allocation of quota rights. If there is a perception of inequality, that should not be blamed on the system, but on its implementation."

He says you can give the quota to a cooperative or a coastal community with restrictions on trading those rights away— albeit with an inevitable loss of economic efficiency. If that is not feasible for some reason, you can tax the companies who do own the quota to provide a social security system for the poor, or to provide training so they can get more skilled jobs. You can only expect so much from the fishery, and the poor are always with us.

In Alaska, where the halibut fishery was once run under the effort controls, the year's quota was sometimes fished out in twenty-four hours. Now the halibut fishery is run under a rights-based system, which allows fishing to go on at a low level all year. It is much more profitable, produces better-quality halibut, kills fewer fish in the process, and gets around the "belly-frying" problem very neatly. The quota is attached to the boat owner, and he has to be on the boat when it is fishing. He may fry his belly if he likes, but he won't get that warm for much of the year in the Bering Sea or the Gulf of Alaska. There is also a rigorous restriction on the vessels that are licensed, which makes it more difficult for there to be technological creep.

Despite the demonstrable advantages to conservation of giving fishermen rights to fish stocks, there is profound suspicion of rights-based solutions among environmental groups. Ironically enough, in the United States, the citadel of capitalism, there is a huge defense of common ownership going on by people who seem unversed in the total failure of common ownership almost everywhere. Even Daniel Pauly, the academic who found that the world's wild fisheries were failing and who sits on the board of at least two environmental groups, is opposed to selling off rights to fish because, he says, they are public assets that already have an owner—the public. He favors instead an auction of rights every year to the highest bidder. I can't understand for the life of me why that wouldn't lead to the same race you get with effort controls, to catch as much fish as possible in the time available—but it's the only point on which we disagree.

I tend to agree more with Rashid Sumaila, a fisheries economist in Daniel Pauly's unit at the University of British Columbia. He does not see why the ownership of the sea or its assets should not be managed as private property any more than the land. Ownership of land confers a peculiar blend of rights and responsibilities under most jurisdictions. One may own land, but one has a responsibility for it. An owner may not pollute his land and may be subject to all sorts of environmental laws and designations as he goes about his business.

Sumaila is a pragmatist. The test, he says, is whether rights would improve the fishery. In Norway the problem is one of policing and trust and proper enforcement. Ownership of rights, by giving fishermen an interest in enforcement, would improve the fishery. In Senegal, he says, social scientists would allege that even if you allocated quotas initially to everyone with a *pirogue*, people would sell their quotas too low because of poverty. The benefits are less clear, the dangers more obvious, but the problem might go away if the quotas were owned by co-operatives.

Whether environmentalists like it or not, the idea of rights to fish seems to be sweeping the world. If they deliver benefits for fishermen and for fish conservation, which they seem to, I think we should be in favor of them. John Gruver, a former skipper with the Alaskan pollock fleet, now working for United Catcher Boats, a small-boats association based in Seattle, drew a parallel between the U.S. prairies and the North Pacific: "In the 1880s they invented barbed wire and divided up the ranges. Now everybody accepts that's the way it is. Nobody likes to see the death of the cowboy. Nobody will think it should be any other way with the sea in twenty years' time."

So far I hope I have shown that the problems in the world's fisheries are more extensive than is often imagined. Most of the management solutions are controversial and complicated, and will always be vulnerable to an overly generous assessment of the stocks. There is one solution that is controversial with fishermen, particularly with those who think they own rights to fish, but it is simple and totally effective. That is preventing fish from being exposed to any kind of fishing gear at all.

DON'T FEED THE FISH

Goat Island Marine Reserve, Leigh, New Zealand. The first national holiday of spring. My half-sister Gillian, a New Zealander for more than thirty years, collected me off an overnight flight from Los Angeles and drove me the 55 miles north of Auckland to Leigh. As we began the hour-and-a-half drive, the Pacific morning air was chilly but clear. The sun was bright and increasingly warm as we negotiated the green hills, the indented coastline, and its mangrove-lined estuaries—the first of many scenic surprises for any traveler who expects all temperate zones to look the same. Suddenly, as we followed the prominent signs to Goat Island Reserve off the highway and drove toward the sea, I could see why an adviser to the fisheries minister had assured me that this national holiday was actually the best time to visit the reserve. The place was crawling with people.

Around a car park on a grassy hillside there were people renting out wetsuits and flippers of all sizes. In front of Goat Island, a tree-covered hillock 180 yards offshore, were people of all ages swimming or wandering about on the sandy beach and rocky terraces. Parties of teenage beginner divers arrived with their instructors as the water began to warm up. Groups of tourists were embarking in a glass-bottomed boat, the most comfortable way of seeing the explosion of fish and underwater vegetation that has taken place since the reserve was created in 1975. At the end of the track, in front of the buildings of Auckland University Marine Laboratory, two figures were waiting for us: Thelma Wilson, a field officer for the department of conservation, a suntanned, uniformed outdoorsy type in shorts, and Dr. Bill Ballantine, godfather to the world's marine reserves, a wiry man in a green marine lab sweatshirt, smoking one of many hand-rolled cigarettes.

Leigh is perhaps the world's best example of what natural ecosystems look like when they are left alone, without fishing pressure, for a long time. The reserve has exceeded its founders' original expectations, and not just from an ecological point of view. Before it was earmarked as a reserve, Goat Island was a popular fishing spot for fishermen. Conflict arose with the marine lab because, as Wilson put it, "people kept eating the experiments." The reserve was created, after twelve years of lobbying by Ballantine and his colleagues at the university marine lab, for narrowly scientific reasons. Ballantine is the first to admit that he never foresaw what an attraction the reserve would become. As we chatted in the sun, he pointed to some of the younger visitors among the five hundred or so present that day. "During all the years we were seeking arguments for doing this, I never heard anyone say schoolkids," he told me grinning. "The reserve came first and the people came after—nobody expected them."

We stood looking at these unexpected beneficiaries of science. Most were families who had driven out of Auckland for the day to stand in the clear sea water and gawk as a profusion of fish swam around them. Others came to snorkel, renting a wetsuit on the beach for a few dollars, or to stare through the glass panel of the boat at the fearless shoals of large snapper, blue maomao, and spotties that swam above forests of kelp, only yards from the shore. It is the only place I know of where you will find signs saying "Please do not feed the fish." These were introduced by Wilson in response to an unexpected problem. What do people do when they find unaccustomed numbers of wildlife in a picnic spot? They offer them a sandwich, of course. The signs explain that feeding can make fish sick, can teach them to bite, and otherwise modify their behavior. There is plenty of natural food for them in the ocean.

What happened to the ecosystem was also unexpected. Snapper are the most prized sporting fish on this coast—and for that reason increasingly small and scarce. In the 1,370 acres of the reserve, the largest snapper are eight times the size of the snapper outside. They are also fourteen times more numerous. Brochures tell you that you will also find butterfly perch, silver drummer, porae, red moki, leather jacket, blue cod, red cod, goat fish, hiwihiwi, butterfish, marble fish, red-banded perch, and demoiselles—all swimming around without fear of people and within a few yards of the shore. Indeed, they have so little fear that they may nibble you to see if you are edible.

Most of this marine menagerie is readily visible to an inexperienced eleven-year-old snorkeler in a rented wetsuit—popular with parents because a child dressed in one tends to float. Further out, in deeper water accessible to more adventurous divers, are delicate gorgonian fans, lace corals, sponges, sea squirts, and anemones. Under the kelp forests, hidden in holes and crevices in the rock ledges, are big rock lobster, or crayfish, much larger than those in the commercially fished waters outside. A line of orange floats stretching out to sea, easily mistaken for the boundary of the reserve, turns out to belong to lobster pots legally set there by local fishermen. Although excluded from the reserve itself on pain of a fine amounting to over $30,000, fishermen are happy because they catch as many rock lobster on the reserve boundary as they would on many more miles of coastline.

When the reserve was created, explains Ballantine, there was nothing particularly special about the marine ecology. The rocky coastal reefs were known as "rock barrens" because nothing grew on them. The most common bottom-living species were large sea urchins, which graze on kelp. As in other parts of New Zealand's northeast coast, the kelp forests had virtually disappeared by the 1960s. The connection between the disappearance of the kelp and overfishing became clear only when the Goat Island Reserve had been established for many years. At a certain point, the undisturbed snapper and crayfish reached a size at which they could prey on the *kina*, or large sea urchins, that fed on kelp. So the kelp forests gradually returned, bringing in turn food and shelter for many other species of fish and shellfish. Biologists call this a "trophic cascade," when the recovery of predators at the top of the food chain has effects that flow down to lower levels. Some of the changes happened quickly. Others took much longer. Some are still happening. It took only three years, in a much larger reserve set up around the Poor Knights Islands to the north, for the snapper population to recover to a level eight times more numerous than before a no-take zone was imposed. The full recovery of kelp forests at Leigh has taken twenty-five years.

The public's interest in the reserve, almost negligible at first, has grown as the transformation of the marine habitat has become more extraordinary. In ten years Wilson has noticed more family groups coming for the day out. Nobody needs an aquarium when they have a reserve like this. The reserve has changed people's idea of what is interesting in the sea. "When we started none of the students would ever dive here—it was boring," said Ballantine. They all wanted to go to the Poor Knights Islands, three hours' drive to the north and 12 miles out. Now students are bobbing about all over the place. "Thelma's one fear," chuckled Ballantine, "is that a shark swims into the bay and she's got to explain to everyone that it's protected."

Like many people who have fostered a big idea that is sweeping the world, Ballantine tends to be regarded with indifference or suspicion in his native land. Now that his convictions have borne such fruit, he is prone to a certain inflexibility of attitude. He can be a curmudgeon, refusing to answer questions if he doesn't like the way they're asked, flaying questioners who don't believe him—recognizable signs, perhaps, of a lifetime spent making a case to people who disagree with him. In his retirement, he is allowing himself the luxury of not being polite to the people he disagrees with. Whether this is wise is another matter.

Now the New Zealand government has introduced a parliamentary bill that would, among other things, allow the setting up of marine nature reserves in a shorter period than the ten years it has taken, on average, to establish each one of those it has so far. "We may have eighteen reserves," said Ballantine, "but all *crept* through." Now the bill is before parliament, and Ballantine is the target of the fishermen again, getting a regular verbal bashing for his opinions and giving as good as he gets in return. I suspect this is the explanation for his take-it-or-leave-it attitude. The fishermen fear him, though, because they know his message is popular. To the conservationists, he is a guru.

Creating a network of marine reserves is obviously an important thing for its own sake. But I was determined to get Ballantine to talk about the potential benefits for managing fisheries, as this is key to getting them accepted politically. And how practicable would it be to manage, say, the North Sea using closed areas alone—as Han Lindeboom suggested all those years ago when I first walked into the wrong lecture in The Hague in 1990? But Ballantine was being grumpy and didn't feel like answering the question, even though he has written papers on the subject. He said it is the wrong question. "Whether reserves would help fisheries is not the point." The point, he said, is that there should be areas of the sea where we see what nature is like in the absence of human transformation. The people who talk about sustainable fisheries, he says, do not really know at what level the ecosystem should be sustained. "One possibility is that you are merely sustaining the wreckage. It could be twice as good, or ten times as good. Without reserves, the check is not possible." I realized I had been trying to stop him from making a really good point.

Okay, I said, and decided to go with the flow because Ballantine was on a roll, focused now on the prevailing official mindset that fishermen are entitled to hunt everywhere in the sea. "We're still assuming that you have one management plan for the whole sea. There is no biological, economic, or social theory that supports this. At no time did anyone say it is a good idea to let everybody fish everywhere."

He told me that at a public meeting he rounded on one sport fisherman opposed to reserves with this question: "Let me get this right: you are saying you want more than 90 percent of the ocean for your own amusement?" History does not relate if the sport fisherman managed to splutter a reply, but it sounds as if, in tennis terms, Ballantine had served an ace. People talk about fishermen being "stakeholders" in the sea, said Ballantine, but that ignores the fact that fishermen may be foreclosing all sorts of options for future generations. "My grandson," noted Ballantine, whose daughter married and divorced a fisherman, "is the stakeholder."

We are back to the old question: who owns the sea? Fishermen like to think it is them—especially when they have been given or sold property rights to exploit a sustainable proportion of the stocks, as they have in New Zealand. But the real answer to who owns the sea is everyone and no one: if there is an owner of a common resource in a democracy, it is the people. Citizens have, until now, had few ways of exercising any influence over what happens in the sea. The voice of the citizen is seldom heard over the voices of the "user groups"—commercial fishermen, sport fishermen, fishing associations, and lobbyists—and establishment fisheries scientists, another interest group that does not necessarily have the same interests as the public.

What people such as Ballantine have done—and it is a very considerable thing—is to circumvent one of the most intractable problems of the oceans, which is that fish don't vote and fishermen do, so politicians go on giving in to fishermen. He has done this by talking to the public over the fishermen's heads, telling them that they own the sea. It is powerful stuff. "I can go into a room full of businessmen, housewives, and Boy Scouts and say we have to keep some of the sea for itself, and it is only the managers and the scientists who are against it. The public are on side. People say, 'We naturally assumed you would be doing that.'" Ballantine has never been more aware of how powerful his message is than at a time when the government is proposing to beef up legislation on reserves—and encountering fierce opposition. "I am finding myself in the position of telling politicians how to buy votes. You can get popular with a million voters by declaring a reserve, and unpopular with 500," he said.

Ballantine began to notice evidence of a shift in the usual pattern of inertia when public opinion stopped Shell from dumping the Brent Spar, a redundant oil platform, in the North Atlantic in 1995. As he wrote at the time:

> Slowly but steadily the public is starting to say—We don't *have* to chop *all* the trees that could profitably be made into useful things. We don't *have* to mine *all* the land that contains useful minerals. We don't *have* to dump rubbish in *all* the available spaces because it would be cheaper. Very soon they will be saying—We don't *have* to fish *all* the sea just because it would be fun or profitable. They are already doing so in New Zealand and the idea is likely to spread.

By now it already has.

Attitudes change. "A hundred years ago people did not go into the bush without a dog and a gun, otherwise people wanted to know what you were doing," said Ballantine. But attitudes have changed more on land than they have so far in the oceans. The advantage of reserves, such as Leigh, is that they enable people to look below the veil of the waves and see fish as wild animals, not as fillets on a slab.

One begins to see why Ballantine's ideas are so revolutionary and why fishermen find his ideas so frightening. No-take zones may not have happened in many places yet, but the idea seems at times to be sweeping the planet—a transformation as big in its way as feminism. Ballantine actually thinks of the idea in this way, likening the change in our view of the ocean to attitudes to votes for women. Why are there numerous marine reserves in New Zealand, yet so few in the mainland United States and virtually none in Europe? He pointed out: "Revolutionary ideas get adopted in small ways miles from the center of established thinking. The first place there were votes for women was in Wyoming territory. One of the last places to give votes to women was Switzerland." A hundred years on from women getting the vote, students cannot comprehend the arguments used at the time for not giving it to them. "If you were organizing a democracy in which half the human race was excluded today, which half would you exclude?" he asked with a grin.

Ballantine has a point: the way things were organized was not necessarily right, proper, or fair but merely reflected the status quo, which people felt must be right. Women had a hell of a job organizing things differently, but they did it. Gradually, attitudes about the oceans are changing, too.

I wasn't going to let Ballantine get away with providing no answer to the question of what benefits reserves have for fisheries. As we were talking, I found what I wanted in literature put out by the New Zealand Department of Conservation. I have described already how snapper on reserves get bigger and more plentiful. Bigger fish produce more offspring. So, of course, do more fish. The snapper egg production on a healthy reserve is therefore vastly greater than in the sea outside. Snapper are estimated to produce eighteen times more eggs on reserves than in other parts of the sea. In fact, snapper egg production on 5 miles of marine reserve is equivalent to egg production on 90 miles of unprotected coastline. Surely this means that there is potential for exporting larvae to supplement and replenish the sea elsewhere, if only enough large reserves were created.

The argument raised against that idea by conventional fisheries scientists—even the conservation-minded Sidney Holt—is that fish move around. You would therefore need reserves that covered vast areas, including entire migration routes, to protect them. In fact, snapper move around, too, and the received scientific wisdom said an increase in the snapper population on New Zealand's marine reserves wouldn't happen. People thought the snapper that got big in the reserve would simply go out and get caught. In fact, it doesn't happen like that, although studies carried out at Leigh and elsewhere show that half the population does take part in a seasonal migration.

If reserves work for snapper, you might ask, could you design a reserve for a highly migratory species, such as the bluefin tuna? Ballantine replies that the usefulness of reserves much smaller than a species' range in conserving its general population has been demonstrated for over a hundred years on land—for birds. The first bird reserves were established around the 1850s. These generally protect only one stopping-off point of a bird's migration, but no one complains that this means they will be ineffective. Scientists Ransom Myers and Boris Worm have found global "diversity hot spots"—Serengetis of the ocean—where migratory species, such as tuna, turtles, marlin, and sharks, congregate. These, they say, would make ideal marine reserves.

So if reserves are a good idea, where should you put them and how big do they need to be? (Ninety percent of the literature on marine reserves, says Ballantine, is about what they would or wouldn't do for target species.) The fact is, he says, that choosing reserves because you want to protect this or that is a mistake. He says the fundamental thing is to protect some of the marine environment. We know too little about it to be sure what the effects of protecting it will be, so his view is to just get on with it and set up a few areas, preferably with some close to where people live. "The idea that we know where to put them is worrying," he said. I know what he means, but it seems a lot easier to have a reason. It makes more sense to tell a fisherman, "We are designating this area because we *believe* on the best possible evidence that it used to be enormously rich and could be again, with many spin-offs for fishermen, if there were no fishing." Should one listen to the fishermen when designating a reserve? Ballantine said: "If we want to do the right thing, we just do it. We wouldn't take any notice if it were, for instance, the construction industry on land. We would just do it." Perhaps fishermen aren't treated just like any other industry often enough.

How much of the sea would you need, then, if you really wanted to conserve fish and used reserves as a deliberate way of boosting the world's fish production? Ballantine didn't have to think for long. "For science and education, 10 percent. For proper conservation of species, 20 percent. For the general good of fishing 30 percent, but if the sea were to be intensively used, you need 50 percent. You need another bay completely no-take for each one that is used." As he spoke I was thinking, what sea deserves the description "intensively used" more than the North Sea? Looking at the tiny area currently devoted to no-take zones underlines the scale of the task involved in creating more. Daniel Pauly has published research saying that if you wanted to take the world's fish stocks back to where they were in the 1970s—never mind the last century—you would need 20 percent of the oceans.

It is arguable that good fisheries management by conventional means is actually unachievable. So many, like Pauly, would argue that reserves, probably quite large reserves, are important not only in themselves but for fishermen. From what we know about fish, these sanctuaries could become repositories of large, superfertile mother fish, laying large, healthy eggs that can restock the oceans.

There is certainly a lot of emerging evidence (published by WWF) that reserves can improve fishing. The Soufrière Marine Management Area in St. Lucia, West Indies, increased local catches by 46 percent from traps and by 90 percent in fishing grounds around four reserves in five years. One of the oldest examples of a reserve that had a tremendous effect in conserving fish is the Merritt Island National Wildlife Refuge in Florida, established in 1962, when it became the security zone for the Kennedy Space Center at Cape Canaveral. After nine years of protection from fishing, the reserve began to contribute world-record-size fish to the surrounding recreational fishery.

An extensive set of reserves, comparable to those in New Zealand, is in the Bahamas. In 1986, the Bahamas National Trust persuaded the government to create a 176-square-mile no-take zone covering the northern end of the Exuma Cays, a long coral reef that occasionally breaks the surface to create a chain of sparsely populated atolls. As a result, tourism has increased and with it local employment. Grouper and the two most valuable shellfish, conch and langouste, have multiplied from a low baseline. The prevailing currents carry their spawn into areas outside the protected area, so everyone appears to benefit. Indeed, so popular was the experiment that the inhabitants of the largest island in the archipelago, Andros, asked for a reserve to be created there, and in 2002 one was declared covering the northern part of Andros and some of the poor, outlying islands. The total area covered by reserves is now 656 square miles. As one conservationist put it: "In my experience, this is a rare instance of pressure for conservation exerted by local people on low incomes."

One should also mention, or course, the 8,500 square miles of sea closed to commercial fishermen off New England, which has resulted in a resurgence of sea scallops, not to mention haddock.

The evidence of reserves producing fisheries-enhancing effects around New Zealand, however, is equivocal. Leigh is one of the few reserves to produce a measurable spillover effect with its spiny lobsters. Experimental fishing around Long Island–Kokomohua Marine Reserve found that blue cod had risen there by 300 percent over seven years, but they remained constant 1 mile or more from the boundary. It may be that the species is so sedentary that any effects will be felt as the export of larvae. This is extremely difficult to measure.

All the same, there is enough evidence that reserves or no-take zones could improve fisheries to make you think that fishermen would be open-minded about them, right? Wrong.

In New Zealand, where the Labour party's Helen Clark leads one of the greenest governments in some time, the Marine Reserves Bill has its back to the wall on the creation of more reserves. At the Department of Conservation (separate from the Department of the Environment, which, for some reason, is not that supportive on marine issues), a huddle of officials had stacked together every possible scientific paper for me on why marine reserves were good things, expecting the same skeptical reception from me as they get at home. In fact, I was already convinced from what I had seen.

Ranged against their bill are, in order of political difficulty, Maori, the fishing industry, and recreational fishermen. Maori and the fishing industry object to the reserves essentially for the same reason—they are big shareholders in commercial fishing operations. I had a tense lunch with Daryl Sykes, who runs the formidable spiny lobster fishermen's association. I happened to remark that it was probably inevitable that there would be a massive expansion of no-take zones all over the world, since this was a commitment signed by 180 or so nations at the Johannesburg summit in 2002. "You want to create a whole load of marine zoos?" asked Sykes. "Why?" The temperature of our conversation leaped to combustion point.

As far as Sykes was concerned, this wasn't a done deal at all. He was a big man, in head-to-toe denim and cowboy boots, and he sat next to me on a bench seat, penning me in and stabbing his fingers at me for emphasis. "There will need to be compensation for loss of access and fishing opportunity," he growled. Rock lobster quota now trades at $143,700 a ton. Rock lobsters do live in rocky ledges along the coast in the sort of places that end up being marine reserves, so Sykes had a point, though how good a point was arguable. You could just as well say that the fishermen were given their quota, so why should they be compensated for small reductions in the area they could fish? The quota system, in any case, only confers the right to harvest a sustainable quota within other constraints, including marine reserves.

The other person present at our tense lunch was Max Hetherington, who represented the recreational fishermen. Hetherington, a chain smoker in his sixties who had already had one heart attack, was a determined spear fisherman, despite the evident risks that diving posed to his red-haired frame. He pointed out that recreational fishermen proposed the first three marine reserves in the South Island, but now they feared a proliferation of reserves up the sensitive coastline—with some good reason. One reserve has been proposed only half a mile along the coast from an existing one. The fishermen had started to fight. As a sport fisherman myself, in fresh water, I found it difficult to understand how the New Zealand government had managed to fall out with the recreational fishermen, who should be the natural allies of marine reserves. It all boiled down to the fact that the Department of Conservation didn't believe in anything short of a no-take zone, having spent years with partial restrictions on fishing around the Poor Knights Islands, which had little effect.

This kind of absolutism, which may well have come from Ballantine, seemed bound to cause trouble because it was seen as an erosion of ordinary people's rights. I remarked to Hetherington that what the Department of Conservation was doing was actually contrary to terrestrial practice. You would never create a terrestrial national park that way in Africa or India. There would have to be buffer zones where there could be some sustainable logging, some hunting, some settlement, as there is in the great rain forest national park at Korup in Cameroon, for example. Buffer zones are now thought to be an essential tool for enlisting the support of local people—but why only in Africa and India and not New Zealand? It seemed to me that a few areas where commercial fishing was banned but recreational fishing was not could have bought crucial support for conservation.

Back in the United Kingdom, I was reminded just how far ahead New Zealand already is in protecting the sea. One of the places I had long thought there should be a reserve was in Lamlash Bay on the Isle of Arran—the place, you may remember, where the best-known sea angling festival in Scotland was once held. In the 1960s, competitors from all over Britain would expect to catch a total of around 7.7 tons of fish over three days. The festival last took place in 1997, when the total catch was just 28 pounds of fish. Lamlash Bay is not unlike Goat Island Marine Reserve, having a large hill rising out of the bay, called the Holy Isle, which is now owned by Tibetan monks and was once a retreat for Christian holy men. The area is very picturesque and has always been a stop on Glaswegians' day trips "doon the watter," when they travel down the Clyde. The last remnant of this nineteenth-century custom takes place in an ancient paddle steamer, the *Waverley,* which still steams through Lamlash Bay to Campbeltown, where Tommy Finn's fishing operation is based. Arran is the Glaswegians' vacation island, and a marine reserve would do wonders for the island's economy. The wildlife there, which included shrimp, tiny flounder, and cuttlefish in the shallows, eider ducks billing at each other in spring, wheeling gannets diving at mackerel in the summer, and dogfish and cod all year round, always made me think that Lamlash Bay would be the perfect place for a protected area when we spent our summer vacations there with my wife's parents, and I nurtured an ambition to propose it one day.

On my return to England, I received an e-mail saying that someone had actually done so. I was delighted and wrote off in return to the proposer, Howard Wood, offering any help I could provide. An organization called the Community of Arran Seabed Trust (COAST) wanted to establish a small no-take zone from which no marine life could be removed, to allow nature to regenerate within the area. They also wanted to declare the rest of the bay as a marine protected area, restricting

clam dredging and commercial diving for shellfish. This would
be the first no-take zone in Scotland and the first community-
proposed one in Britain. In England a very small no-take zone
has been set up, by the government agency English Nature,
around the long-established marine reserve (where fishing was
still allowed) at Lundy Island in the English Channel. The
Lundy one was top-down. The Lamlash one would be bottom-
up, the result of people power. It germinated out of a meeting
of COAST's chairman, Don Macneish, a decade earlier with—
you guessed—Bill Ballantine.

The second time I got in touch with Howard Wood things
weren't going so well. COAST now had the support of five hundred people—a quarter of the population of the island—but
they had had a meeting with Gabriella Pieraccini, the devolved
Scottish Executive's head of inshore fisheries. She had said that
the campaigners would have to get agreement from the Clyde
fishermen before the project could be completed. The Clyde
Fishermen's Association was opposed to the idea. So much for
the people owning the sea. Wood, a diver, said COAST had received some support from Scottish Natural Heritage, the statu-
tory conservation body, when they had pointed out that part of
the seabed was an old, wrecked maerl (calcified seaweed) bed
thousands of years old, a favorite nursery area for fish and shellfish. The maerl bed would recover if it were left alone. The bay
itself had long been protected, as it was used by the navy for submarine maneuvers and as a mooring place for larger vessels, but
now the shoals of queenies (small scallops) were being fished
out fast. Wood remembered that these queenies used to rise
from the seabed like clouds of butterflies when you passed by.
Recently he found a clutch of a dozen together, more than he
had seen for years.

Howard Wood said the impression he had been given was

that the Scottish Executive—which happens to be at the center of a complaint from the European Union for not applying the law with sufficient rigor to stop fishermen from cheating—was trying its best not to upset the fishermen. Meanwhile, two scallop dredgers had arrived in Whiting Bay, just a short distance along the coast, and had been scraping the bed of the bay for scallops all week within a few yards of the shore. Local people, he said, were outraged that this was legal and that they could do nothing about it. The Clyde Fishermen's Association told them that what happened on the seabed off Arran's shores had nothing to do with its residents.

The idea of leaving parts of the sea alone is very simple. It cuts across the ideas of traditional scientific fisheries "management," with its impressive-sounding professionals telling us how much they know. Scientific fisheries management has failed almost everywhere, sometimes because of some factor that scientists failed to predict, sometimes because politicians wouldn't listen to them. The beauty of large reserves for biodiversity and fish management is that they are an insurance policy against this kind of failure and a reminder of how marine ecosystems behave in the absence of human transformation. It is obvious from experience that they work. Monitoring them with satellite technology is a piece of cake. You make fishermen fit their vessels with transceivers. If you find a fishing boat has been in a reserve without any extenuating circumstances, you take away its license to fish or imprison the skipper.

I find it reassuring that simply protecting some parts of the sea for their own sake and the sake of the fish is being seen by more and more scientists who understand all the complexities of so-called fisheries management as the way ahead, because it gets around incompetence, political egregiousness, and intellectual dishonesty by their own kind.

A ROD TO BEAT THEM WITH

A few months after writing the British version of this book I went back to New Zealand with my family on vacation. On the way back from a bit of trout fishing and a relaxing stay on the beach, we visited the Goat Island reserve and stayed with Bill Ballantine. Over dinner he and a friend and my wife, Pamela, and I had a passionate discussion about recreational fishing. Was it a force for conservation or a threat to the marine environment? I made the case for the ethical, conservation-minded angling. Bill's friend, who worked for the Department of Conservation, was equally forceful in arguing that there could be no compromise with recreational fishing when setting aside areas of sea as marine reserves. The next morning, after the empty wine bottles were cleared away, I decided that a discussion about angling needed to be included in this book. For I had to concede that the Kiwi conservationists' basic premise was right: you can't let anglers off the hook when it comes to overfishing just because they use a rod and line.

From the Kiwi's perspective, recreational fishing was barely distinguishable from commercial fishing. Commercial sport fishing is the biggest use of the sea along the North Island coast—as it is in Florida and along the Gulf coast of the United States. We had just been staying at a place owned by a family of avid sea anglers. There were piles of fishing magazines with pictures on the cover that made my small sons gasp: smiling men holding very large dead fish, including marlin, yellowtail kingfish, and the overfished snapper. Some people call these mags fish porn, not without justification. Inside there were advertisements for rods and reels that were able to control a running fish better than ever, navigational equipment, boats, and fish finders. You could see that it was not just in commercial fishing that technological creep had accelerated to technological gallop.

There was clearly quite a difference between all this subtropical, seagoing tackle and the trout kit I had in the trunk of my car: a few old fly reels, sold to last a lifetime in my father's day, and some new carbon-fiber rods. But was it really one of more than degree? Fly fishing can be a very technical business on big, clear New Zealand rivers, as it is in the Rockies. We had been releasing our catch anyway, but the bigger fish had a habit of releasing themselves—complete with my fly in their mouths. You could argue that this was cruel, though it was inadvertent. I find it hard to accept the equivalence proposed by the animal welfare group PETA, which once devised an advertisement with the image of a spaniel with a giant tuna hook through its lip as an attack on angling for sport. The argument becomes about the physiology of fishes' mouths, about pain, and about consciousness. My practical observation is that fish with small flies in their mouths seem to behave pretty normally, and eventually the flies come out. The same would not be true of an escaped fish spiked by an assembly of treble hooks or bait fishing tackle, which would be fatal.

These inhumanities pale into insignificance, however, compared with the multiple atrocities of a 50-mile long line with thousands of baited hooks indiscriminately catching turtles as well as fish, trawls catching thousands of fish at a time, or the netting of some relatively intelligent creatures such as octopus. I am uncomfortable drawing any kind of absolute moral equivalence between a human, or a mammal, and a fish. I recognize that is an important area of philosophical and scientific inquiry, but ethical questions about the treatment of the individual, rather than the conservation of the species, stretch far beyond my abilities to resolve them and, more importantly, beyond the scope of this book. Meanwhile, the ethical commercial fisherman tries to let the fish he doesn't want escape and I try, where possible, to use barbless hooks.

What we do know with reasonable certainty is that overfishing happened first in fresh water and with low-tech methods. Much nineteenth-century debate on both sides of the Atlantic in the 1860s and 1870s was about why salmon stocks had declined. We know that overfishing can be caused by angling with rod and line, as studies of Atlantic salmon runs conducted in the 1980s showed that the spring run—the fish that come into the river early on the big spring floods and wait until autumn to spawn—can be fished out by angling pressure alone.

In April 1982 I caught an unusually large salmon in the River Dee, in Wales. It weighed 23 pounds. It was my first salmon after five years trying to catch one. So I banged it on the head, carried it home in triumph, and we ate it smoked for months. Every year until the mid-1980s there had been a few large spring salmon like this one caught on the Dee. From then on, the large fish stopped coming, and the spring run virtually died out. It was too long before the authorities banned the killing of spring fish. By then the damage had been done. There were many other reasons why the spring run failed: the cutting off of spawning streams in the hills, pollution from industry and agriculture, drift netting at sea, legal fishing in the estuary, and poaching. But the number of anglers and the deadliness of the technology they used—spinner and prawn being easier to use and more effective than fly—was a factor in the decline of the spring run. My small part in the decline of the spring salmon got me thinking about what else was going on in the river and at sea. It was the genesis of this book.

Now I rejoice to say that more action is being taken on the Dee and the nets at the river mouth have been bought off, so there is a chance that the spring run will come back. It did, as a result of deliberate reintroductions, after an equally dire period caused by pollution and poaching before the First World War.

The traditional picture of the angler as someone catching a few fish for his supper needs to be modified. At sea, sonar, fish finders, and global positioning systems have improved so much that they can locate shoals and even large individual fish. Small, fast, easily transportable boats make it easier to get to sea and back. The skipper is doing it for a living, and the results of the expedition are taken home for the freezer. The culture of this kind of sport fishing is often more in line with the viewpoint of commercial fishermen, who ultimately believes the sea is a giant larder. There is potentially a difference in kind between this sort of fishing and that of the freshwater fishermen, who fishes in season for enjoyment with a tiny fly that imitates nature and who releases most if not all of his catch. The century-old ethics of sport—which, after all, mean giving your quarry a fair chance—arguably have not kept up with technology any more than have the ethics of commercial fishing.

So it is not very surprising that a study of recreational salt-water fishing in the United States over the past twenty-two years found that up to a quarter of overfished species were taken by recreational fishermen, not commercial ones. The study, published in *Science* in August 2005, found that 60 percent of red snapper in the Gulf of Mexico were taken by sport fishers. Felicia Coleman, the lead author, from Florida State University, Tallahassee, told me: "The conventional wisdom is that recreational fishing is a small proportion of the total take, so it is largely overlooked. But if you remove the fish caught and used for fish sticks and fish meal, the recreational take rises to 10 percent nationally. And if you focus in on the populations identified by the federal government as species of concern, it rises to 23 percent." Coleman says there are now 10 million saltwater recreational anglers in the United States, and the sport has grown 20 percent in the past twenty years. So there are too many saltwater anglers for present conservation measures to work.

Coleman observed that state licenses, which cap the number of fish you can bring in, were unlimited in number. Commercial fishermen had an annual quota, and when it was reached the fishery was closed for the year. But anglers, though limited to the number of fish they could keep per trip, had no quota and went on fishing. This was a genuine problem. Either the number of fish that fishermen could take needed to be limited or the number of licenses did. It does also seem questionable how much angling pressure should be permitted on stocks that are vulnerable. Disturbingly, the amount of cod taken in the Gulf of Maine by the recreational fishery now amounts to around 30 percent of the catch.

As Carl Safina, author, conservationist, and president of the

Blue Ocean Institute, a nonprofit body based in Cold Spring Harbor, New York, said when Coleman and her colleagues' study came out: "There's little use in commercial and recreational fishers pointing fingers at each other. Commercial fishing is not all bad and recreational fishing is not all good. A fish doesn't care if you are a commercial or a recreational fishermen. It only cares if it is surrounded by water—or on ice."

That is true, so far as it goes, but it doesn't go quite far enough. Commercial and recreational fishermen will always be in competition with each other, and society has to have some yardstick for deciding on the merits of their competing claims. The background to Coleman's paper is that some species such as red snapper are now limited to angling only in Florida because of the damage that was done to them before by commercial methods. Enlightened restrictions, coupled with sport fishing, have contributed to the restoration of species such as red drum, snook, speckled trout, bluefish, and flounder. As fish get scarce and anglers more plentiful, there has to be an evaluation of what is the best economic, environmental, and social use of the fish, and hard decisions need to be made about which method of capturing them has the best potential for allowing depleted stocks to recover. The answer to that is the more selective the better.

There are few places in the Northern Hemisphere where the value of fish stocks for recreation, leisure, and the seaside economy better understood than in the eastern United States. The restoration of the striped bass from North Carolina to Maine after a collapse in stocks in the 1970s ranks as the most successful restoration of a fin fish stock in North America. The striped bass is now enjoyed as a sporting fish by millions of people. It is instructive to compare the remarkable value in terms of the citizens' pleasure as well as the money that is brought into the local economy by the striped bass's resurgence with the sorry mismanagement of its cousin, the European species of bass.

Fishermen on the East Coast of the United States have delighted in the size and fighting abilities of the striped bass since the earliest colonial times. The "striper" spawns in fresh water but spends most of its life in the ocean. It takes five to seven years for spawning females to reach full productivity, so the population is vulnerable if too many large fish are caught. A fifteen-year-old bass can carry 3 million eggs compared with a six-year-old's half a million. What happened was that a commercial catch of 14.7 million pounds in 1972 dropped to 1.7 million pounds ten years later. Sport fishermen reported an equally severe drop in catches, particularly in Chesapeake Bay, the spawning ground and nursery for nearly 90 percent of the Atlantic population. The consequence of the striped bass's decline was a loss of around 7,000 jobs and $220 million in revenue.

The decline so alarmed Congress that it enacted the Emergency Striped Bass Act of 1979. Researchers from the U.S. Fish and Wildlife Service, state agencies, and universities who carried out a study initiated by that legislation found that because there were fewer larger females, which produced more vigorous young, the bass population was much more susceptible to natural stresses as well as pollution. Acid rain (produced by power station emissions) and aluminum (which the acid leached out of the soil) were fatal to newly hatched larvae. Bass larvae were also affected by toxic pollutants, agricultural pesticides, and the chlorination of effluent from sewage plants. The study concluded that reducing fishing pressure would have an immediate positive effect. A plan devised by the Atlantic States Marine Fisheries Commission set limits on the size and weight of fish that could be retained. Some states went further. Maryland imposed a total moratorium on catching striped bass in 1985, followed by Virginia. Thousands of hatchery-reared bass were introduced to rivers.

By 1989, the bass population had increased. Both Virginia and Maryland lifted their moratoriums, and limited commercial and recreational fishing resumed. Since then anglers have been limited to one fish per day and the minimum size limit of 28 inches ensures that all fish should have reproduced by the time they are caught. Trophy fish of over 40 pounds are back.

As the stocks of striped bass improved, sport fishermen successfully persuaded legislators of the greater socioeconomic benefits of recreational fishing. Some states, such as Maine and Connecticut, now manage the fish entirely for recreation. The recreational fishing industry has grown at a rate beyond the wildest expectations, increasing from about 1 million trips a year in 1981 to more than 7 million by 1996. Commercial landings that year were worth $8 million. Over the years 1981 to 1996, angler expenditure, adjusted for inflation, rose from $85 million to $560 million. Some now estimate it to have reached $2 billion. Anglers fly from as far away as Europe to fish for stripers in hot spots such as Montauk, Long Island, and off Cape Cod.

Recently commercial fishermen have been pressing to have access to higher quotas of striped bass or rockfish, as they are called in the Carolinas. They point out that 76 percent of all striped bass are killed as a result of sport fishing, and that the mortality from catch and release alone (25 percent) exceeds the commercial quota (18 percent of the stock in 2003). With millions of bass being caught and released on millions of fishing expeditions a year, the mortality adds up to a significant number. Even so, the overall value to the millions of citizens who benefit from the fishery and to the East Coast economy is demonstrably higher than a larger commercial fishery would create.

Compare the resurgence of the striped bass with the sorry plight of the European bass, which is largely at the mercy of commercial fishermen. As this book goes to press, the commercial outfits in Britain are still bitterly resisting an increase in the minimum landing size from 14 inches (half the minimum size for striped bass in the United States) to 17¾ inches, which would benefit both commercial and sport fishermen.

The European bass spawns at sea and its eggs are carried back on the currents to estuaries, which become nurseries for the young fish until they are three to five years old, when they migrate to sea. In the 1950s, the bass was primarily a recreational species. Since then its popularity as a culinary fish has led to a directed fishery with gill nets around the coast and pair trawlers in the English Channel and off northern France. Dolphins run with the bass shoals, and the pair trawlers and their two-thirds of a mile of midwater net are thought to be responsible for killing the hundreds of dolphins that wash up dead each year.

Warmer sea temperatures over the past twenty years have produced an explosion of bass around the British coast. There has been a proliferation of small bass in the Thames, Blackwater, and Stour and Orwell estuaries. But with the minimum landing size set by the British government at 14 inches, the maximum size of fish in the estuary is usually about 2 pounds. The bass are graded by the gill nets like peas in a bag.

You would think that there was a conservation reason for choosing the present minimum landing size of 14 inches. The reason turns out to be that it corresponds with the size of bass the market traditionally prefers, the size of a plate. It doesn't have anything to do with conserving bass. Bass don't usually spawn until they are at least 17 inches, so most bass sold in the little markets around the coast by fishermen unlikely to pay tax on their landings will not have reproduced. In two years of fishing there, we have only caught one larger fish on rod and line. My son Harry, then age twelve, caught a bass of 4½ pounds on a ragworm. Otherwise we content ourselves with putting back all the fish of a pound or two and hoping that enough larger ones exist somewhere out at sea and will make it past the nets to spawn.

So does sport fishing have a role in restraining the overfishing of the oceans or will it always be part of the problem? As Carl Safina observed, there is no inherent difference to the fish between a load of individual fishermen catching a lot of fish and a single trawler catching the same number. There is, however, a value in having anglers involved, because managing fish is really a matter of managing fishermen. By and large, anglers' culture is more precautionary. They are more likely to advocate stinting to achieve their ends than are fishermen who sell their fish for a living because anglers want to catch larger fish that have spawned more often.

Commercial fishermen know this. At a recent meeting of ICCAT, the Atlantic tuna conservation body, representatives from the United States suggested that recreational interests should be represented when discussing the worsening state of bluefin stocks in the Mediterranean, as they are when discussing the future of the bluefin on the eastern seaboard of the United States. There was a brief discussion among officials from the European Commission, which is overwhelmingly influenced by commercial interests, before the idea was emphatically rejected. I doubt that decision was made in the interests of the bluefin.

McMEALS FOREVER

Click on McDonald's Web site and you may learn more than you wish to know about what goes into a 5-ounce Filet-O-Fish. That is, unless you want to know about the fish itself. You are told that the battered fish in the bun is pollock or hoki. McDonald's does not explain that the pollock is walleye pollock from Alaska and that hoki is one of the cod family from New Zealand, or what the proportion of hoki is to pollock. Nor does it tell you how either was caught or whether the fishery is well managed. McDonald's "nutritional facts for your McMeal" assume you want to know that the batter on the outside of the fish is made from bleached white flour, water, modified corn starch, yellow corn flour, salt, whey, dextrose, disodium pyrophosphate, sodium pyrophosphate, sodium tripolyphosphate, and cellulose gum—presumably in case you have an allergic reaction to one of these ingredients and need to seek medical help. You are left with the impression that McDonald's Web site is written for potentially litigious, nutrition-obsessed neurotics with a strange lack of curiosity about the provenance of what they are eating.

I can't help being curious, though, about what McDonald's

means by a "filet." It is sometimes referred to as a "fish filet patty," which raises suspicions that what we are dealing with here is minced fish (which it isn't). Minced fish is what makes up the lower grades of processed white fish in the fish fingers and breaded shapes sold in supermarkets. There is actually nothing wrong with it. Sometimes minced fish has a higher fat content and is therefore superior in taste. Americans prefer their fish not to taste too fishy. Europeans prefer a fishier taste. So fish for the United States market will have what processors call the "fat lines," the lateral lines of darker-colored meat, taken out. This also improves freezer life because the fat lines can go rancid even when frozen. Fish fillets for the American market are called "deep-skin fillets," having been skinned to leave just pearly white flesh.

After talking to McDonald's and some of the people in New Zealand, Seattle, and Alaska who supply the company, I can reveal that what is actually in a Filet-O-Fish is a deep-skin fillet, but not necessarily what the average person might think of as a fillet. McDonald's, which is known in the processing trade for having such demanding standards that some processors find the company too taxing to deal with, uses fillets frozen into blocks. These are frozen at sea or in an onshore processing plant in Alaska. At the secondary processors, where they go to be battered, they are sawed into fillet shapes while still frozen, so each Filet-O-Fish may contain parts of several actual fish. The fish "fillet" thaws a bit during secondary processing when the warm batter is applied, but is immediately refrozen. It is then sent on to restaurants, where it is cooked. Compared with the fish sold in ordinary restaurants, which is iced but not frozen at sea and might have been out of the water for three weeks before you eat it, McDonald's fish is actually very fresh, having usually been out of the sea only a matter of minutes before it is frozen. The freezer life of a pollock block for most food manufacturers is a year after primary processing. (Freezer life depends on fat, which depends on species. The 100-year-old orange roughy, for example, is virtually indestructible in the freezer.) Strangely, the McDonald's Web site doesn't tell you any of this.

None of this is leading where you might expect, and certainly not where McDonald's might expect a book such as this to lead. For both the fish used by McDonald's in the United States, Alaskan pollock and New Zealand hoki, are certified by the Marine Stewardship Council (original slogan "Fish forever," now "The best environmental choice in seafood"), an organization that gives an independent certification of sustainability, an ecolabel, to fisheries. McDonald's uses Alaskan pollock in 90 percent of the 275 million fish sandwiches it sells in the United States and Canada each year. The most surprising thing is that McDonald's is making so little marketing capital out of their use of two out of the three white fish stocks in the world that are actually certified as sustainably caught. Or at least it surprised me until I found out that McDonald's would have to pay royalties for the use of the MSC label.

The very idea that McDonald's, hated by greens and foodies, could make capital out of being on the side of the angels—and that their customers were therefore more virtuous than the denizens of exclusive restaurants—caused shivers of revulsion among the righteous in U.S. environmental groups. When the Alaskan pollock fishery proposed itself for certification as one of the world's best-run fisheries, they accused it, in the words of one campaigner, of "strip-mining the ocean and treating the fish like a crop of corn in Iowa." The man who said that was Ken Stump, a Seattle-based, hardworking, conviction-driven environmentalist working for Alaska Oceans Network. He could not believe that you could give the world's largest, most commercially successful white fish fishery a pat on the back for being ecologically sound. Put that way, neither could I. So I booked a stopover in Seattle, the home of the Bering Sea and Gulf of Alaska pollock fishery, to check it out.

The idea of harnessing consumer choice to build a world of better-managed fisheries seems to have occurred quite independently to one of the world's largest buyers of fish, Unilever, and to the world's largest conservation organization, the World Wildlife Fund, in 1995. Mike Sutton, a U.S. citizen working for WWF International and based near London, concluded that the world's wild fish stocks would not last more than a few decades unless drastic action was taken. He reckoned WWF was never going to get the action it wanted by traditional advocacy, that is, lobbying governments, a process he likened to herding cats. He discovered by chance that Unilever, the largest fish buyer in the West and owner of such brands as Birds Eye and Knorr, was also becoming alarmed at the extent of overfishing, which it considered a serious risk to its frozen fish business. Mike arranged lunch with Simon Bryceson, a consultant to Unilever, at the Groucho Club in London to compare notes and plot.

There they discussed creating an alliance between business and environmentalists to certify fisheries, thereby giving the consumer the power to choose fish from independently certified well-managed stocks. The idea wasn't original—a Forest Stewardship Council had already been established with the intention of combating similar problems of greed and poor governance that led to the felling of primary forests. But persuading the fishing industry to trust itself to an alliance with environmentalists was an entirely different kettle of fish.

There is nothing wrong with the idea of using certification to establish and improve standards. Certification has existed for over a century and, broadly, it works. Teachers, doctors, nurses, police officers, auditors, engineers, ships' pilots, and lab assistants require certification to practice their trade. Environmental certification goes back to 1946, when the organic farming movement was born. Certification companies, like firms of auditors, exist all over the world and are used to buying in expertise. The only trouble Unilever and the WWF appear not to have foreseen was that the people who staff certification companies tend to communicate in impenetrably dull technical language that the consumer may not find terribly reassuring.

Simon Bryceson, at Unilever, persuaded Catherine Whitfield, a native of maritime Canada who had experienced the cod crisis at close hand, that working with WWF was in the corporation's interest. Anthony Burgmans, a tall Dutchman on Unilever's board, who later became co-chairman of the company, bought the idea. He signed an agreement with WWF at a ceremony in The Hague. Not long after, Mike Sutton phoned me to say that a new organization was to be formed called the Marine Stewardship Council, which would organize the independent auditing of fisheries. We journalists sold the story to our editors by saying that governments would come out of this badly by comparison with good practice around the world, because at the time scientists who worked for the government were virtually the only judges of what was happening in fisheries.

Unilever made an unusual promise for a major transnational company, vowing in 1996 that it would be sourcing all its fish from sources certified as sustainable by 2005. The company warned in 2003 that it would not meet its target, reflecting the slow progress in converting the world's fisheries to sustainable management practices. Nevertheless, by the end of 2005 it had sourced 56 percent of its fish from what it considered to be sustainable sources, some 49 percent from MSC-certified fisheries. In the process it had given a jolt to the system of all the fisheries it dealt with. Early on the company asked all its suppliers to confirm that their fish was legally caught and that they were not involved in species threatened with extinction. It claims to have stopped doing business with those suppliers who could not offer this kind of confirmation. It does not buy cod from New England or the North Sea—though you could argue that these are not able to provide much volume anyway—and has sold its tuna-canning business. It also asked the fisheries it deals with, apart from Alaskan pollock—Norwegian cod and saithe, Chilean and South African hake—to seek MSC certification. South African hake was duly certified after Alaskan pollock. That, a pollock fisherman in Seattle told me admiringly, is what you call leadership. As far as I can see, he was right.

Ten years on, the now-independent Marine Stewardship Council—WWF and Unilever withdrew once the organization was set up and running—allows fifteen fisheries to carry its blue ecosymbol, among them western Australian rock lobster, Thames herring, Alaskan salmon, New Zealand hoki, South Georgia toothfish, Baja California spiny lobster, South African hake, Gulf of Alaska and Bering Sea pollock, and Alaskan long-lined cod. A long list of fisheries—presently topped by North Sea herring, at 300,000 tons—is undergoing assessment. There is a precertification phase, which is not made public, in which no-hopers are privately told they stand no chance, to save them embarrassment. If all the fish awaiting certification do get certified, 6 percent of the world's fish supplies will be sustainably managed according to the MSC's rules. That may not sound like much, but the MSC is already an influence in important markets. Some 32 percent of the world's supply of prime white fish (cod, saithe, pollock, hake, haddock) is now in an MSC program, as are 42 percent of all wild-caught salmon and 18 percent of spiny lobster. More importantly, the MSC took a major step toward being the only global ecolabel for fish—which was what its founders wanted it to be—when the UN Food and Agriculture Organization issued guidelines on ecolabeling in 2005 that said certification bodies had to be independent, science-based, and inclusive of all shades of opinion. That meant anyone setting up a comparable certification scheme would have to go through the same hurdles that have taken the MSC ten years already, which is highly unlikely. The risk that the wild-capture fishing industry will go it alone and set up its own labeling schemes to greenwash its behavior—which has happened in fur farming—has suddenly become remote.

Yet just as the MSC stood on the verge of a breakthrough with the certification of the world's biggest remaining white fish fishery, in Alaska, it ran into trouble. There were plenty of initiatives competing to harness consumer buying power in the United States, and some environmental groups were suspicious of industry and didn't really see the point of the MSC. Excellent seafood guides showing customers what to buy and not to buy had been produced by the Monterey Bay Aquarium, the National Audubon Society, and the Blue Ocean Institute. There were also several fish ecolabels in the United States, one of which, Eco-Fish, was in trouble on the front page of the *Seattle Post-Intelligencer* the day I arrived to check out the pollock fishery. A chain of natural food stores had been using its label, erroneously, on tuna.

There are cultural differences between environmental groups in Europe and the United States, which didn't help either. Where European environmental groups have focused on ecolabeling as a guide to personal choice, those in the United States have a strong history of achieving things through organized boycotts. SeaWeb, a nonprofit organization based in Washington, D.C., and the Natural Resources Defense Council boycotted the Atlantic swordfish, then called it off when they perceived a stiffening of resolve to reduce quotas and tackle illegal fishing by ICCAT. The National Environmental Trust organized a campaign to persuade consumers to "take a pass on Chilean sea bass." Gerry Leape, the organizer of that campaign, was one of the most prominent critics of the MSC—first from outside, then as a member of their stakeholder council. He began by asking some extremely good questions.

How do you decide whether a fishery is sustainable? Can any fishery conducted by 270-foot trawlers, as is the Alaskan pollock fishery, ever be said to be sustainable? That was also Ken Stump's question, if you remember, and a good one. What is the acceptable degree of interference in the ecosystem? The MSC has established three principles it believes a sustainable fishery is based upon. It then requires certification companies retained by applicants to consider the fishery against those three principles. A fishery needs to score 80 percent or more under each.

Principle 1. A fishery must be conducted in a manner that does not lead to overfishing or depletion of the exploited populations, and for those populations that are depleted must be conducted in a way that demonstrably leads to their recovery.

Principle 2. Fishing operations should allow for the maintenance of the structure, productivity, function, and diversity of the ecosystem (including the habitat and associated dependant ecologically related species) on which the fishery depends.

Principle 3. The fishery is subject to an effective management system that respects local, national, and international laws and standards, and incorporates institutional and operational frameworks that require use of the resource to be responsible and sustainable.

The process cannot have seemed difficult when the MSC certified the Thames/Blackwater herring, a smaller version of the North Sea herring, which spawns between the rivers Thames and Blackwater in the east of England and is caught in small drift nets and landed at West Mersea on the Essex coast, not far from where I live. A total allowable catch of 128 tons is permitted each year. The MSC certifiers met in the Company Shed, a rough-and-ready seafood café on the front in West Mersea, and placed a single condition on the certification: that this inshore fishery find a way of coordinating all its vessels to ensure that fishing ended when the quota was used up. That was a piece of cake.

Little difficulty can have been involved either with the Loch Torridon langoustine creel fishery in the northwest of Scotland (there is a distinctly British flavor to the London-based MSC's earlier certifications, as it was cutting its teeth). The fishermen use fixed, baited traps, a vastly less damaging method than trawling. Creel fishermen found themselves in conflict with trawlers because—as the inhabitants of Arran have also discovered—Scotland removed a ban on fishing with mobile gear within 3 miles of the shore in 1984. The Loch Torridon fishermen applied to the Scottish Executive to have an area closed to mobile gear. They succeeded in 2001, when a closed area was established in Loch Torridon and the Sound of Rona, which is likely to bring many benefits, and not only for langoustines. The creel fishermen make a good income catching about 100 tons of langoustines a year, all of which are exported to Spain.

It was when the MSC certifiers got to hoki, one of the world's major commercial fisheries, that things got complicated. The fishery is carried out by 250-foot freezer trawlers in the Cook Strait between the north and south islands of New Zealand and in the Pacific and the Tasman Sea. The fishery is conducted under individual transferable quotas, with draconian penalities for cheating, including the confiscation of quota, vessels, and licenses to fish. There was a question about how much the stock had been fished in the past and how long it should be allowed to recover. There was originally an annual bycatch of 1,000 fur seals and 1,110 seabirds, over 60 percent of the latter being species of endangered albatross. An objection was lodged by a New Zealand environmental group, the Royal Forest and Bird Protection Society. It said that the certifiers had failed adequately to interpret and comply with the MSC's three principles. It was the first certification to be vigorously opposed.

The MSC set up an independent panel to consider whether a certificate should have been given. This said that the way the fishery was run was "world best practice" and confirmed its certification. But the panel was sympathetic to Forest and Bird's view that little had been done to reduce the bycatch of seals and birds. The panel recommended that seal excluder devices should be tried. The MSC says its approach is "one of continuous improvement, and it is not necessary that all problems be solved prior to certification." Since the review panel in 2001, the annual audits the MSC requires from its independent certifiers have criticized the Hoki Fishery Management Company for failing to complete sea trials of excluder devices. It did not look good when the hoki company had to ask the New Zealand fisheries minister in 2003 to cut its quota because the hoki stock appeared to be in decline. Forest and Bird appears to have had a point that quotas were set too high and the stock may not have been managed for recovery, in accordance with the MSC's Principle 1. If the hoki stock declines further, the MSC will have some explaining to do. On the other hand, the seal deaths were down to 150 last season, the number of young fish seems now to be growing, and seabirds are now seen as less of a problem than in other fisheries.

You can see why the proposal to ecolabel South Georgia toothfish might have alarmed Gerry Leape, organizer of the "Take a pass on Chilean sea bass" campaign. But competition between unsustainable supplies and sustainable ones was exactly what the MSC was supposed to create. In time a legal, sustainably managed fishery, provided its fish can be clearly identified by a chain of custody (shortly to be approved), could squeeze out illegal supplies in major markets at least. Ecolabeling is therefore more of a sharp instrument than a boycott is. What matters is that the fishery is well run and well policed and that the stock is discrete, that is, not part of a wider stock that is being poached.

An afternoon at Imperial College, London, with David Agnew, who designed the management system for the South Georgia government, convinced me that South Georgia toothfish are confined to the continental shelf around the island. The other areas where toothfish are found—and poached to hell—are on the other side of a very large continent. The South Georgia fishery is patrolled by the British vessel *Dorada,* which is armed with a heavy machine gun, and surveyed by satellite and overflights from the Falklands. An illegal vessel was fired on and sunk last year. Around South Georgia the albatross and petrel bycatch is down to twenty a year because long-liners are required to set their lines under deterrent streamers and at night.

The Alaskan pollock raised the greatest number of questions despite the management of the fishery being, in the words of the assessors who spent two and a half years reviewing it, world-class. The reason is the complicated part the walleye pollock plays in one of the world's last functioning semiwilderness ecosystems. Between 1970 and 1998 the pollock fishery in the North Pacific and the Bering Sea contributed about 4–7 million tons a year to the world's supplies of fish. Half of this fishery, in the Bering Sea and the Gulf of Alaska, belongs to the United States, the other half to Russia, where there is thought to be significant overfishing. In U.S. waters, catches are running at a staggering 2.2 million tons a year, all of which is caught by U.S. vessels. Since 1977 there seems to have been a major regime shift in Alaskan waters. Pollock numbers have soared and the spawning stock is now as large as it has ever been. However, Steller sea lions, an endangered species covered by the U.S. Endangered Species Act, are 80 percent down since the 1970s. Scientists and environmentalists blame the decline on fishing too close to the coasts, where pollock, a main source of food for the sea lions, are found. Other scientists blame killer whales, which used to attack great whales and may have moved on to tasty young sea lions.

I visited the offices of the At-Sea Processors Association in Seattle, which was behind the application for pollock certification. At-Sea Processors, which represents the large freezer trawlers, is unlike any other fishing company I have ever visited. The building is bustling with lawyers in suits. It is clearly a very profitable operation. I spoke to Craig Cross, who worked for a company running 270-foot freezer trawlers—pollock fishing is a midwater trawl with very little disturbance of the bottom—and he explained the agreement in 1999 that allowed fishermen to form cooperatives that divided up the quota. It worked in a similar way to a rights-based system. He said it had eliminated the race for fish, making the fishery much less ecologically wasteful and much more economically efficient. "We'll spend a day looking for the right size of fish, where in the past you just kept fishing," said Craig. All large trawlers carry observers. John Gruver of United Catcher Boats, which represents smaller vessels, around 120 feet long, that land their catch in Alaska, said that in the absence of a race to fish, fishermen were now competitive about bycatch avoidance. There is a large bycatch of salmon, which native Alaskans depend on for food, in the pollock fishery. The rules say you have to stop fishing when you reach your limit of salmon. "The fisherman of the future isn't going to be measured by the fish he does catch, but by the fish he doesn't catch," said Gruver. I went away thinking that these were the most sophisticated and enlightened fishermen I had met anywhere.

Then I met Ken Stump of Alaska Oceans Network and heard about the dark side of the fishery. There was a plausible link between sea lion decline and overfishing in the Shelikov Strait in the Gulf of Alaska. There were no marine reserves in the Bering Sea or the Gulf of Alaska—though there were substantial closed areas. The fishermen had also not mentioned the winter fishery for 550,000 tons of spawning pollock to provide roe for the lucrative Japanese market. These pollock, if left to spawn, would contribute to a much healthier population of fish. So this did sound like strip-mining the oceans after all.

The only point on which I disagreed with Ken was his view that the MSC was a "promotional gimmick of big industry players and had little to do with protecting the marine environment." It seemed to me that this charge remained to be proved. There was also the question of the influence on other fisheries that certifying Alaskan pollock would have. The United States imported 70 percent of its fish, two-thirds of it farmed, much of the rest from fisheries less well managed than the Alaskan pollock fishery. What other solution to the problem of making fisheries more sustainable was he offering? I made a speech to a Seafood Choices Alliance conference in Seattle saying that if some U.S. environmentalists thought the Alaska pollock fishery was badly run, they should get out more.

The certifiers eventually scored the pollock fishery over 80 percent against all three MSC principles. The MSC awarded it a certification, subject to fairly stringent conditions. These, if applied, will drag the pollock fishery some decades into the future and be a victory, ironically, for Ken Stump. Studies will have to be carried out to demonstrate that the industry's harvesting strategy is precautionary even in the event of regime shifts, the fishery will have to take ecosystem considerations into account when planning where fishing effort should take place; it will also have to improve assessments of the impact of fishing on the foraging habitat of the Steller sea lion. The fishery has to comply for the first time in twenty years with two domestic laws, one of them the Endangered Species Act. But (a black mark against the certifiers) there was no requirement to look into the establishment of marine reserves or protected areas.

A criticism of the MSC is that it is impossible for a lay person coming cold to the organization's Web site to make head or tail of whether it has done its job properly. At least a quarter of what the independent certifiers have written is opaque jargon that has no business in a public document. Large parts of it are fisheries science gibberish. Its competence is neither evident nor persuasive. It has to be taken on trust, which was precisely the approach of so much of the government science and policy that has left the oceans in their current sorry state. It is not written to give consumers confidence in the fish they eat, only to convince a paymaster, which happens to be the industry, that all the *i*'s have been dotted and *t*'s crossed.

I was so annoyed by the reports on the pollock fishery that I thought I would do a little journalistic digging to see if I could find out anything they didn't know that would puncture the pomposity of their professional certifiers. Sure enough, I did discover something that didn't seem too sustainable. In Alaska the waste from the vast amount of pollock that is processed to make all those McDonald's fillets and Unilever fish fingers gets boiled down to make oil, together with waste from other fish that get caught with them and the "trash fish" that nobody has a use for. This oil, amounting to some 33,000 tons a year, could easily be used in human fish oil supplements, in food for farmed fish, or to feed animals, thereby reducing the amount of small fish (such as the overfished blue whiting) caught globally for that purpose. Instead, the oil is burned to generate electricity.

The main reason for this absurdity is the Jones Act, a piece of protectionist U.S. legislation dating from 1924. This says that any vessel sailing between two American ports must have an American crew, something that makes transport between Seattle and Alaska vastly expensive—and makes fuel expensive, too, for it has to be shipped with American crews from Seattle, not the low-paid multinational crews now commonplace on merchant ships. This was told me by Scott Smiley of the University of Alaska's fisheries technology center in Kodiak, who added: "It's a double win for them because of the transportation cost of fuel from Seattle." Until the fuel is taxed or the Jones Act repealed, this obscenity is likely to continue.

I pointed this out to a member of the MSC in Seattle, who agreed it was "ludicrous" but added that what happened to waste was not generally part of any certification. In my view that was pure sophistry. Remember MSC Principle 3, insisting on "institutional and operational frameworks that require use of the resource to be responsible and sustainable." Burning fish oil, measured against any criteria you like, isn't responsible or sustainable, for reasons I discuss in the following chapter.

So should the pollock fishery have gotten an MSC certificate? It comes down to two questions. How good is the fishery? And—dear old Rashid Sumaila's killer question—what would improve it? The answer to the first was about 7 out of 10. With a little effort it could be an 8—good enough. What would improve it? What would mean, for example, that there was more chance of something being done about the Alaskan fish oil scandal? To me the answer was clear: certify the fishery. If you thought, as some environmentalists appeared to, that we should only certify perfection, what was going to spur improvement? It would have been an act of despair not to certify the pollock fishery, warts and all. Now we are waiting for the warts to be removed.

Two and half years after that visit to Seattle, the MSC is seen quite differently in the United States. Alaskan pollock is now on the green, do-eat list of the Blue Ocean Institute and the Monterey Bay Aquarium. The MSC has done a breakthrough deal with Wal-Mart, the biggest retailer in the world, which says it is now seeking to source all its wild-caught fish and frozen fish within three to five years from MSC-certified sustainable fisheries. (There remains scepticism about this promise among Wal-Mart's critics.) The MSC champions also include Whole Foods Market, Compass, Metro, and Safeway. Rupert Howes, MSC's chief executive and a qualified accountant, says the Wal-Mart deal will bring pressure to bear on less well-managed fisheries, even the notorious Russian side of the Bering Sea, which Wal-Mart also buys from, to come under assessment. The only thing that is proving difficult for Howes is tackling those companies that could use the MSC logo, such as McDonald's, but are too cheap to pay for the privilege.

So, back to the beginning: should my conscience be troubling me if I eat a McDonald's Filet-O-Fish? Those phosphates with long names do trouble me, and so does the calorie count. It bothers me they don't use the MSC logo. But if we are talking about what is going on in the world and how to make a difference, then I would advise the stick-thin patrons of exclusive restaurants selling endangered species to walk out and get their skinny asses to McDonald's.

BURNING THE MIDNIGHT OIL

Third Fisherman: Master, I marvel how the fishes live in the Sea.
First Fisherman: Why as men do a-land, the great ones eat up the little ones.
　　　　—William Shakespeare, *Pericles*, Act 2, Scene 1

Esbjerg, the largest port on Denmark's North Sea coast. A musty smell used to waft into the town square on most days, particularly when the wind was coming from the direction of the giant 999 fish meal plant down on the docks. Workers at the factory used to lower a large vacuum pipe into the trawlers' holds to suck up the silvery sand eels, sprats, or pout. Billions of little fish were then carried into a massive industrial process that squashed, mashed, and boiled them, then divided them into oil and fish meal. The meal was blended with other supplements to make food pellets for farmed pigs, chicken, and salmon. The oil went partly into the pellets for salmon, the rest to make human food supplements, margarine, paints, varnishes, and other products.

The first time I went to Esbjerg, fifteen years ago, it was to investigate rumors that the trawlers supplying the plant were catching so many fish that there had been a glut of fish oil. The factory manager, a diligent fellow, had found a new market for his surplus of oil: he was selling it to power stations. The oil had nearly the same energy value as fuel oil, with the added advantage that it was not taxed. It made poor-quality coal burn extremely well.

In Denmark, where they take a particularly utilitarian view of the sea, only a minority of people were alarmed by the practice. In Britain and over most of Europe the burning of fish oil was regarded as an abomination. Denmark's fisheries minister was forced to slap a tax on fish oil used as fuel to persuade the 999 factory to stop selling it to power stations.

The sand eel, known as the sand lance in the United States, is the fish that nearly everything else eats. It lives in the mud for half its life, and the factors driving its extraordinary fecundity have never been satisfactorily explained. My story about what the Danes were doing caused an outcry among salmon and sea trout anglers in Scotland, where catches were in decline. It upset bird organizations concerned about the number of unexplained bird deaths around the North Sea. Scottish trawlermen were stirred, too, because they have long accused the Danes of stealing the food of the cod and haddock and vacuuming up juvenile herring, haddock, and cod as bycatch in their fine-mesh nets. Those who catch white fish also knew that if the North Sea's cod and haddock were to grow large again, the sand eel fishery would have to be reduced. Its huge abundance is thought to result from the overfishing of its predators.

Now is a dispiriting time for the 999 plant and the trawlers that supply it. Last year the entire sand eel fishery was closed because sand eel catches dropped by a third. The numbers of individual fish were half the 300 billion that the European Commission says are the minimum in the sea for it to continue. Workers have been laid off, and the only way the plant has stayed open is by processing foreign landings of blue whiting. The 999 factory was also hit early in 2005 by a drop in sales after researchers found high levels of cancer-causing dioxins and PCBs (chlorinated compounds) in the salmon in Scottish, Faroese, and Norwegian fish farms. The contaminants were concentrated in the salmon food made in the 999 plant by boiling down small fish from Europe's polluted waters. The PCB levels were such that some scientists recommended eating only one portion of farmed salmon a month.

The pollution dates from the days, only two decades or so ago, when the North Sea was treated by industry as an open sewer and a dumping ground for toxic chemicals. Although the dumping of PCBs is now banned, the discovery of contamination in salmon flesh cast a cloud over fish farming, which many "experts" still favor as the answer to feeding an expanding human population and to solving the problem of declining wild fish catches. By 2030, says the UN Food and Agriculture Organization, fish farming will dominate fish supplies. Given how wrong the FAO has been in the past—saying catches were going up when, in fact, they were going down—this statement is worth examining carefully. When you do, you find it to be an observation of previous trends, not a reflection of what could happen or what people might want—in the same way as Red Delicious was once far and away the most popular apple in the United States because it was basically the *only* apple you could get. The FAO is simply observing that fish farming is the fastest growing form of food production in the world—growing at 9 percent a year and by 12–13 percent in the United States. Nobody is asking us whether we want this. It is just happening. The continued destruction of mangrove swamps in poor countries to provide shrimp for people living in rich countries is simply the market operating in a vacuum untroubled by ethics. It is a reflection of what will go on happening if we do not find ways of exercising any choice in the matter.

I am not opposed in principle, as many environmentalists appear to be, to fish farming, even to an industrial process dependent on factories such as the 999 plant in Esbjerg, if it can be run within sustainable limits. I happen to think that if I was hungry enough, I would be happy to eat farmed fish every day. If I were poor enough, I might well rip down a mangrove in Vietnam or central America to farm tiger prawns for export, assuming I knew the right cocktail of chemicals to keep them alive, which I gather is difficult. But it is the responsibility of those of us who are fortunate enough to exercise choice to ask whether this is the way the world should be going.

I think that we, as consumers of fish, have to start with what we want, then modify this with the reality of what we are likely to need. What we want, all the trends show, is more fish, preferably tasty shellfish, such as big tiger prawns, or oily fish low in mercury and PCB contaminants and containing plenty of omega-3 fatty acids. What we need, if there are no wild fish or they are too expensive, is farmed fish. We seem less troubled by fish farming that damages the environment—perhaps we see it as a necessary price to pay—but this may also be because we know very little about it. On the few occasions that people are asked, they say they would like fish farming to be sustainable, but this may be difficult. Let us examine the trends in fish farming, then the constraining factors, and try to see where we, as consumers, have any influence.

There are two kinds of fish farming, or aquaculture, as it's called by jargonwriters and people who want to neutralize the negative connotations of intensive farming in the sea. The first was developed by the Chinese at least two thousand years ago and entails feeding waste vegetables to fish in a pond. The fish are herbivores or omnivores, chosen for their tolerance to fairly murky water. When people talk about the growth of low-tech fish farming for subsistence in developing countries, they are talking about this. There are fewer problems with keeping vegetarian fish, such as tilapia. Domesticated strains grow faster than the wild varieties. It is an option the developed world should consider doing more. The problem is one of taste. The taste for carp, for example, has pretty much died out in western Europe, but not in central European countries, such as Poland and Hungary. Medieval monks imported carp into Britain from the Continent and kept them in ponds so that they had fish to eat on Fridays. All the same, to western European tastes, saddle of carp is pretty inedible.

What the West thinks of as fish farming, however, is the kind that has grown up in the past thirty years, which relies on feeding processed food made of ground-up wild fish to carnivorous fish, such as salmon, trout, and prawns. The growth figures have been little short of staggering—shrimp, mostly produced in the world's poor South, is now the most popular seafood in the United States. (I have never seen a food writer mention this, but all shrimp imported into the United States must first be washed in chlorine bleach to kill bugs. What this does for the taste, I do not know, but I think we should be told.) Salmon is now the third most popular seafood in the United States. A decade ago Costco, one of the biggest retailers, did not stock fresh fish. Now it sells 16,500 tons of farmed salmon fillets a year. The cost of mistakes, mainly disease caused by keeping wild animals in close proximity, but also escapes during storms and overexpansion, has meant that not everyone has made their fortune out of fish farming.

The whole carnivorous fish industry depends on the continued availability of small, wild fish, such as sand eel, menhaden, capelin, and blue whiting, to grind up into pellets. That is why there is trouble on the horizon, for the amount of small wild fish is finite, and so is the amount of fish oil and meal. Of the two, the determining factor is fish oil, which contains all the omega-3 fatty acids that we and salmon thrive on. This is because 70 percent of what is currently produced is already used in fish food. On the other hand, only about 34 percent of fish meal currently goes into fish farms; 29 percent goes into pigs, 27 percent into poultry, and 10 percent into various other human and animal foods. Dr. Stuart Barlow, director general of the International Fishmeal and Oil Organization, told me he warned the fish feed industry seven years ago that fish oil would be a limit to growth. He cautioned that if they didn't find ways of substituting vegetable oil for fish oil, and to a lesser extent vegetable protein for fish protein, the world would be unable to answer any new demands for fish food by the end of this decade.

Remarkably, the amount of fish oil and meal produced was static from 1950 until the end of the 1990s. There are hiccups in supply every time there is a major El Niño warming event in the Pacific, when the reproduction of the Peruvian anchoveta tends to drop off. The anchoveta is the subject of the biggest industrial fishery in the world—and was in the 1970s the subject of a major population crash. In the future, as the price rises, it is thought that the use of fish meal and oil for purposes other than fish farming will drop away, and virtually all fish oil will go into fish feed. Barlow says that if people were concerned to cut waste, they should take their fish oil in capsules rather than it going back into fish food. Provided the supplier can assure you that the PCBs and dioxins have been taken out, that is an alternative to eating farmed salmon—a bit like cutting out the middleman.

So is fish farming the solution to the decline of wild fish in the oceans and the search for more human food, or is it just a new problem? In theory, fish farming could hold the solutions to some of the problems facing wild fish. It is unquestionably the solution to the possible extinction of bluefin tuna in Europe caused by uncontrolled catches for the fattening cages. The Japanese have closed the circle with bluefin tuna by breeding it in captivity, but this is much more expensive than catching it in the sea, which needs to be banned or heavily regulated by the Convention on International Trade in Endangered Species. If the Japanese and smart restaurateurs like bluefin tuna so much, let them pay for rearing them. The Chileans are reported to have closed the circle on the Patagonian toothfish, so poaching toothfish may be about to go the same way as poaching salmon, a major source of revenue when there were prolific quantities of wild fish.

The key question underlying the sustainability of any farming operation for carnivorous fish is where all the small wild fish that go into the feed are coming from. The static nature of the global industrial catch conceals the fact that we have been fishing down the food chain. Forty years ago we were pulping North Sea herring to make pig food, which contributed to the herring's collapse. Now there is a decline in the sand eel. This means few places are safe for some of the world's other stocks of unloved, bony, unattractive, but ecologically important fish.

We know there are grotesquely unsustainable catches of blue

whiting currently coming from the North Atlantic on vessels such as the *Atlantic Dawn*. If farmed salmon need three times their body weight in food made from other fish, then why don't we just eat the blue whiting? As I found when I asked the British chef Rick Stein to cook blue whiting fillets, they taste pretty good. Blue whiting has far fewer residues of PCBs and dioxins because it has not been boiled down and concentrated and because it comes from the comparatively unpolluted waters of the mid-Atlantic. Blue whiting has absolutely no antibiotic residues. It contains none of the pesticides that are used to kill the parasitic sea lice on the salmon, and it is free of the flesh colorant—banned for use in terrestrial food—that is used to make the salmon's flesh pink. The only difference is that, like cod, it concentrates oil in its liver, not in its flesh. Blue whiting is not an oily fish, so if you want your omega-3s, you would be better off eating herring, which is in reasonable supply.

Horse mackerel, another of the world's staple industrial fish, *is* an oily fish. It is prized by the Japanese and has graced at least one emperor's breakfast table. I have never eaten it, so I can't tell you what it's like, but even sand eels taste nice deep-fried. So what is stopping us eating blue whiting, horse mackerel, sand eels, or even the Peruvian anchoveta? Nothing, except that the market is not yet used to providing them. So instead of asking for farmed salmon the next time you're at the fish counter, try asking for blue whiting or horse mackerel and see what happens. You might just start something.

The blue whiting rush is not the only aspect of industrial fishing that raises ethical concerns. I am indebted to the sharp eyes of Monica Verbeek of the conservation group Seas at Risk for pointing out a disquieting entry in the records of the Danish industrial catch. It appears that 4,195 tons of the round-nosed grenadier, a deep-sea fish, found their way into Danish fish meal plants in 2002. The round-nosed grenadier is a fish that first reproduces at the age of eight to ten, lives to seventy-five years, and is many times more vulnerable to overfishing than the short-lived, highly fecund species that are normally thought of as "industrial" fish. It is conceivable that this large tonnage of grenadiers was accidentally caught while fishing for something else, probably cod or ling, and would otherwise have been thrown over the side. Maybe this catch was merely efficiently used waste that prevented some other fish having to be caught. The alternative is too horrible to contemplate. Directed fishing for deep-sea species for fish meal would represent a descent into a new circle of hell, anarchy, and stupidity for the world's fisheries. I am told that the Irish fleet has considered it.

If fishing for deep-sea fish to boil down is a descent into hell, then there is another circle of hell devoted to a new refinement of industrial fishing. This is what the *Economist* calls "the diversion of low-value fish from the mouths of people in developing countries into the mouths of well-fed fish in the developed world." I have heard of limited examples of this happening already: whole sardines and pilchards from the great Mauritania fishery off West Africa being fed to bluefin tuna in so-called tuna farms in Australia, which fatten tuna for the Japanese market. If this becomes widespread, it would amount to an obscenity on an imperial Roman scale.

If the pressures on the world's small wild fish are like this now, then what will they be like in a few years' time? It is no longer salmon but halibut and cod that are causing a wave of interest in northern and western Europe. Cod are being seen as the answer, in Norway and the EU, to cheap competition from Chile. Research has been poured into the technical problem of how to feed a fry that, unlike salmon, has no large yolk sac it can live off in the early days of its life. The research seems to be paying off. In 2004 the huge fish farming company Nutreco brought 385 tons of cod to market. By 2020 the company expects to be farming 440,000 tons of cod in Norway, mainly for export to the now largely cod-free EU. The figure of 440,000 tons is almost the same as the quota for wild fish caught in the Barents Sea by the entire Norwegian and Russian fleets. I have not yet seen any sign that anyone is thinking of the possible implications of this massive expansion for wild cod. Would fish farms take the pressure off wild stocks, allowing them to recover to their original abundance, or is it more likely that farming would introduce genetic pollution and disease into the wild population? Cod farms tend to use wild-caught herring and mackerel as feed. Would we not be better just eating the herring and mackerel, which are healthier, oily fish?

The proposed massive expansion of cod farming in Europe is nothing, however, compared with what is being contemplated in the United States, where the George W. Bush administration has proposed a huge expansion in offshore fish farming. The proposal, made in summer 2005, has opened a debate in Congress about how to balance the nation's need for fish with a stream of uncertainties about the environmental effects of the rapidly growing industry. Fish farmers want the federal regulatory agencies to privatize parts of the open ocean for fish farms, though how this fits in with wild fish is unclear. Currently there are experimental offshore fish farms situated 3–200 miles from the coast—in Alabama, Mississippi, Florida, Hawaii, New Hampshire, Puerto Rico, and Texas—growing high-value species, such as red drum, amberjack, summer flounder, cod, halibut, red snapper, and cobia. Even assuming that the technology exists to construct them without allowing huge escapes of domesticated fish that could breed with wild fish with unpredictable consequences—and it does not—there doesn't seem yet to have been any consideration of what open-ocean farms might do to the seabed in terms of pollution or to wild populations by passing on new diseases incubated by caged fish. The proposed legislation would give fish farmers the right to use the global commons as their sewer. It would also conflict with proposals to give the catchers of wild fish property rights over the stocks they catch.

Technology, as ever, is neutral and can provide solutions as well as problems. A new system of farming salmon on land, using pumped seawater, claims that it could slash production costs by a quarter. The production centers would be near markets, cutting transportation costs and food miles. The water would be recirculated and purified using bacteria, then sterilized, cutting the 20 percent mortality rate in sea cages, and the waste would become high-value products used in medicines and pharmaceuticals. The beauty of the recirc system, from the point of view of wild fish and those who benefit from them, is that there is little chance of escape of domestic fish likely to infect or genetically pollute their wild relatives. The downside of the recirc farming system is the high establishment costs and the sheer size of the tanks required to compete with sea cages containing 2,750 tons of fish.

There is a conflict between wild fisheries and aquaculture that has a long way to run. In the United States there is talk of using genetic modification to insert a gene into salmon to make it grow bigger and faster. In many other countries there is a belief that genetically farmed fish are a step too far. The concern is about what happens if the fish escape and breed with wild fish, as they are almost certain to do unless they are kept in contained facilities on land—and even then the security of these cannot be guaranteed. Disastrous introductions of alien species have occurred around the world. There was the poisonous cane toad in Australia, introduced to kill pests of sugar cane, which now kills a whole range of native wildlife from snake to crocodiles, not to mention domestic pets, each year. There was the Atlantic jellyfish in the Black Sea, which ate up all the zooplankton and replaced the anchovy, which previously formed a valuable commercial fishery. And now I gather that the Atlantic salmon has been introduced by salmon farmers to the Pacific and, though fish farmers assured us this would never happen, has escaped and begun to breed in the rivers. All these examples stand as reasons why genetically modified fish are not a good idea.

You would think natural, indigenous parasites had caused enough trouble. I will touch only briefly on the cautionary tale of salmon farms on the west coast of Scotland—an example it is worth thinking about carefully before undertaking the kinds of expansion of fish farming that are being contemplated. All my life I have been a sea trout angler, and the west of Scotland should be my mecca. The first time I went there to fish Loch Maree, then the best sea trout loch in Scotland, was in the mid-1970s, when I was a teenager. At that time at least 2,500 sea trout migrated annually up the River Ewe and were caught by visiting anglers in Loch Maree. When I went there fifteen years later, the Loch Maree Hotel, one of the most famous fishing hotels in Scotland, was empty of anglers. There were a few devotees who went out to fish for sea trout, but the annual catch was under a hundred, and one year numbered just nineteen. The same thing has happened to river systems all over the west coast and the Scottish islands because the estuaries and the coast have become infested with huge concentrations of sea lice as a result of salmon farms. Once the presence of a sea louse on the side of a fresh salmon was a cause for celebration—the sign of a fish fresh from the sea, as sea lice drop off in fresh water. Now it is a cause of foreboding.

For nearly a decade I had been writing stories about the mounting evidence that sea lice were responsible for the sea trout collapse that happened in the 1980s, just as salmon farming took a firm hold on the Scottish west coast. Most of the evidence came from Ireland, where scientists, such as Ken Whelan of the Salmon Research Agency, were quicker to report and more open about the problem. In Scotland, where salmon farming went virtually unregulated during its formative years, nobody in the establishment wanted to know anything that might detract from what was seen as a wonderful way of creating employment in rural areas. In this case, it was a form of employment that put others out of business.

Then, all of a sudden, the scientific position on sea lice changed. Professor David Mackay, an area director of the Scottish Environmental Protection Agency, told a gathering of his peers in Norway in 1999: "The case that damage to stocks of sea trout and wild salmon by sea lice associated with caged farming is very serious in certain circumstances has been made to the point that it should now be accepted as beyond reasonable doubt." Three years later, the strongest evidence to date between sea lice infestation of wild migratory trout and fish farms was published by the Scottish Executive's own fisheries laboratory at Faskelly. It demonstrated that the infestation of sea trout in Loch Torridon and the River Sheildaig was of farmed origin, and this was shown to be at its worst (i.e., potentially fatal to wild fish) in the farmed salmon's second year in the pens. WWF said it was the smoking gun that proved that farmers were responsible for destroying the wild fish. Allan Wilson, then the Scottish minister for environment and rural development (a difficult combination of roles to pull off successfully), responded that it was "crucial to strike a balance between the needs of this growing sector and its impact on the environment." Striking a balance in this case meant doing nothing.

Four years later not a single salmon farm has been moved from the path of migratory fish. Scotland's home-grown anti-aquaculture campaigners continue to blow holes in the industry's credibility, and there is increasing pressure on the industry from all sides. Comparable research has been done, too, in Canada, where a study funded by the Pew Charitable Trusts found that infection levels from parasites in wild juvenile salmon near one salmon farm were seventy-three times higher than normal. The science is firm enough now in Scotland, one would have thought, for a court case for damages to be brought by the owner of a fishery against a salmon farm, but none has yet materialized. A welcome development is that some fish farming companies, such as Marine Harvest, have begun to use their undoubted skills in farming fish to restock rivers, such as the Lochy, with sea trout. I shall believe that the fish farmers have earned their place as full members of society only when they restore Loch Maree.

All the same, fish farmers are not going to go away. Like other industries, they place their faith in technological improvement in what is still a very young industry. They make the point that many of the food scares, which both the United States and Europe seem addicted to, are just that. The industry knew about dioxins and PCBs in European waters several years ago, and knew also that the regulators on both sides of the Atlantic were likely to introduce limits for toxic chemicals in fish food. Two years before the researchers' findings about PCB contamination appeared in *Science,* the 999 plant in Esbjerg began treating the fish meal to remove high levels of dioxins and PCBs. Had researchers tested Scottish salmon in 2004, rather than five years ago, the results would have been very different.

Graeme Dear, managing director of Marine Harvest in Scotland, assured me that the fish oil shortage, too, will be solved by innovation. Nutreco, Marine Harvest's parent company, now believes that it can substitute 75 percent of the fish oils in fish feed with vegetable oils without any ill effects for the salmon, or presumably for the person eating it. Ulf Wijkström of the FAO talks about the "fish feed trap" being solved by using "as yet unexploited aquatic proteins," such as Antarctic krill. Ultimately, fish farming may not be that good for us, but it is better than having no fish at all, or fish that is unaffordably expensive. We will have to live with the fish farming industry, and the industry will increasingly have to live with us by paying more of its environmental costs.

Some salmon farmers have tried to deal with society's fears about intensive fish farming by offering organically grown salmon. This is a respectable try, but undermined by the fact that fish have to be fed with concentrated fish-meal that is made up of fish often caught in unsustainable ways and containing, of course, contaminants in varying concentrations depending on where the feed stock is from. An attempt to devise a trustworthy labeling system for the least environmentally damaging farmed fish—such as the organic movement provides in agriculture—is earnestly to be wished for so that the consumer of farmed prawns, salmon, and sea bass has an environmentally friendly choice; also, competition drives up standards. It must be asked, though, how environmentally friendly it really is to farm carnivorous fish that need several times their own weight in other fish to reach maturity. Would we farm tigers? Nobody seems to be asking fundamental questions. I suspect there will be money in it for someone who makes vegetarian fish taste palatable—as they can with tilapia in Israel. So far, the only aquaculture assurance schemes are run by the industry itself. And the only organic fish in most people's minds are wild fish, which are preferable anyway.

The scariest thing is that nobody seems to be considering the impact on those wild fish of fish farming on the scale that is now being proposed on the coast of Norway or in the open ocean off the United States. Fish farming, even with conventional techniques, changes fish within a few generations from an animal like a wild buffalo or a wildebeest to the equivalent of a domestic cow.

Domesticated salmon, after several generations, are fat, listless things that are good at putting on weight, not swimming up fast-moving rivers. When they get into a river and breed with wild fish, they can damage the wild fish's prospects of surviving to reproduce. When domesticated fish breed with wild fish, studies indicate the breeding success initially goes up, then slumps as the genetically different offspring are far less successful at returning to the river. Many of the salmon in Norwegian rivers, which used to have fine runs of unusually large fish, are now of farmed origin. Domesticated salmon are also prone to potentially lethal diseases, such as infectious salmon anemia, which has meant many thousands have had to be quarantined or killed. They are also prone to the parasite *Gyrodactylus salaris*, which has meant that whole river systems in Norway have had to be poisoned with the insecticide rotenone and restocked.

What would happen if there were large cod farms near the largest concentration of wild cod left on Earth, in the Barents Sea? If Norway's regulators have their way, we could be about to find out.

Increasingly, we will be faced with a choice: whether to keep the oceans for wild fish or farmed fish. Farming domesticated species in close proximity with wild fish will mean that domesticated fish always win. Nobody in the world of policy appears to be asking what is best for society, wild fish or farmed fish. And what sort of farmed fish, anyway? Were this question to be asked, and answered honestly, we might find that our interests lay in prioritizing wild fish and making their ecosystems more productive by leaving them alone enough of the time. We might find that society's interests do not lie, as the FAO has led us to believe, in a massive expansion of aquaculture.

On land, incentives have already swung from intensive forms of agriculture toward extensive ones. In England, at least, pigs are no longer raised in concrete silos, but outdoors with room to snout around and wallow in mud. Following the problems caused by bovine spongiform encephalopathy (BSE, better known as mad cow disease), terrestrial animals are no longer fed their own kind mashed up and rendered down. It is the fish farmers' nightmare that a BSE equivalent lurks for farmed fish. There are compelling reasons for favoring extensively grown fish rather than intensively grown ones. But where do you see that in any government policy? Everywhere in the sea the incentives and subsidies favor the killing of too many wild fish and the building of more fish farms.

This is the opposite of what society says it wants. Just as nobody asked whether it was a good idea to take a million tons of sand eels from the North Sea and boil them up in the 999 factory, it just happened. What will happen in the future is what the aquaculture industry thinks will make money, unless we make an active choice to arrange things differently. We do have a choice as to whether the incentives driving the aquaculture industry—unimaginative regulation and often subsidies— should exist. We do have the choice of restructuring wild fisheries so that they are no longer a way of buying the votes of fishing communities but a way of managing fish stocks properly so that they provide health and nutritional benefits to society. We would be healthier and eat fewer chemicals and pesticides concentrated in farmed fish if we were to have a clear hierarchy of priorities, with wild fish at the top and farmed fish at the bottom.

We need to remember that the conservation of wild fish is a human health issue.

THE THEFT OF THE SEA

It is easy to get too depressed about the state of the world's oceans. I think a more creative response is anger, because the problems are not insoluble, though they are certainly very big indeed: On our journey around the world we have seen that catches of wild fish have peaked and are now in decline; that rational management of fisheries is the exception rather than the rule; that the strip-mining of the most valuable fish is altering diets and even evolution; and that the developed world is stealing food from less fortunate countries and curtailing options available to future generations. We have seen that the growth in aquaculture is raising questions rather than providing answers. It is time to begin asking some political questions, starting with one of the most fundamental of all: who owns the sea anyway?

The owner of the sea within 200 miles, in a modern democracy, is us, the citizens. Or at least it should be. The reality across much of the planet is that most of the common benefits that the sea provided before rapid advances in fishing technology are being taken from the citizen almost everywhere by fishermen as they fight it out for the last fish. A celebrated British writer once said that the farmers had stolen the countryside.

Well, this time it is the fishermen's turn. They have stolen the sea. The world over, bureaucrats and politicians alike assume that commercial fishermen are the constituency they have to satisfy, and not the true owners of the sea, the citizens. There are echoes of what George Orwell wrote about another common system, albeit a fictional one, in *Animal Farm:* "Some animals are more equal than others." Yet the paradox of the undermanaged commons is that some fishermen have also stolen livelihoods from other fishermen. The result is a state of affairs that suits very few people indeed. And for *people* you can also read *species*.

In Great Britain I once worked out that the wild capture part of the commercial fishing industry was roughly the size of the lawn mower industry. No one would dream of allowing the lawn mower industry alone to dictate the policies of a sovereign state, or federation, in any other field of human activity. Where this happens the lunatics have truly taken over the asylum.

The fact that the sea is dominated (to a greater extent in the European Union and other countries than in the United States and Australasia) by people who believe culturally that fishing—and now aquaculture—is a justifiable activity in 100 percent of the sea breeds a culture that is corrosive. It leads to a clutch of absurd notions that would not be tolerated in any other food business or in agriculture: the belief that scientists are the enemy, that you can cheat biology, that foreigners fiddle and our fishermen don't, and that you can set up a regulatory system but (wink) rig it to fail and everyone will be happy, because the inexhaustible sea will ultimately take up the burden of all your inability to govern effectively.

The tyranny of the fisherman's point of view leads to information about the seas being primarily presented, where it is presented at all, in a form that is accessible only to fishermen and scientists and not to the rightful owners of those resources themselves. In Spain, one of the most rapacious fishing nations on earth, most fisheries information that does not have to be revealed by law is treated as commercially confidential. It would be nice to think that one day every nation would publish the equivalent of Iceland's Blue Book for all the marine species under its jurisdiction, and include the baseline data for every stock in 1850, 1950, 2000, and so on, so citizens can see just how much of their birthright has been squandered—or, just possibly, how much one day it has been restored.

The cozy culture of listening to the fishermen perpetuates the cultural assumption that you can keep people happy in far-flung communities from Labrador to the west of Ireland and from Gloucester to the rugged Galician coast by allowing them to fish, when in fact the galloping advance of fishing technology means that this year maybe only half a dozen people can fish sustainably with that kind of gear and next year it will be four. Right now twenty of the rogues are at it.

I believe citizens are beginning to realize that their birthright, a healthy ecosystem, has been stolen, and they want it back. And smart fishermen are realizing that their fish have greater value, both social and economic, if they come from a healthy ecosystem. So what can citizens do about getting back their birthright? Citizens can impose ethical pressure on the fishery and on politicians through their buying habits. This is very important. Ultimately, though, good practice needs to be backed up by regulation and by political decision making.

I think it is possible to isolate good political practice from bad, or at least to define the *worst* practice, for I live closer to it than many Americans. It is called the European Union's Common Fisheries Policy.

It is instructive to compare American fisheries—and, come

to that, Australasian fisheries—with European ones because the comparison neatly highlights the difference between bad and good practice. Where American and Australasian fisheries are better run, which in many cases they are, it is because of greater involvement in the fishery by people who are not commercial fishermen, whether they are recreational fishermen, divers, or environmentalists.

As a journalist writing about environmental problems for the past eighteen years, I have spent a fair amount of time reporting criticism of the United States for reacting slowly to the discovery of the destruction of the ozone layer over Antarctica (first predicted by an American scientist, then missed by U.S. space satellites, then found by a British Antarctic scientist with a balloon and a piece of string, and finally acted on by the world community) and for doing the same about climate change. Whatever you may think of the George W. Bush administration's decision to renege on the Kyoto treaty in 2001, the result of the United States' actions has been to allow the European Union to delude itself that it is a world leader on a broad range of environmental issues. The reality is that it has a pretty indifferent record so far on reducing its emissions of greenhouse gases and an awful one on fisheries. On fisheries, the United States, while very far from perfect, has gotten a few things right. Not everyone knows that.

I first realized just how awfully wrong things had gone in Europe's waters while talking to a senior British official in the mid-1990s. I asked him what it felt like to play off the demands of fishermen every year against the needs of the North Sea cod, which has been threatening collapse on a Newfoundland scale. He let the mask of officialdom slip for a moment and admitted it was like backing toward a cliff in a fog.

You know, he said, that the edge is there somewhere. You can hear the sound of the waves crashing and the cries of the gulls wheeling below, but you do not know exactly when you will topple over. I realized then that Europe's fisheries were out of control. If the sound of the waves and the cry of the gulls were loud then, they are much louder now.

Let me briefly summarize the indictment against the European Union for its handling of its common seas and its negligent control of fleets that fish in other oceans. Two-thirds of Europe's main commercial fish stocks, including cod, hake, plaice, and sole, are being harvested unsustainably. Some species, such as the common skate, have virtually died out. Fishing vessels kill dolphins and porpoises at unsustainable rates. The Mediterranean, bordered by twenty-two countries, six of which are in the EU, is the only sea on Earth with no 200-mile limits and few controls on mesh size, so most fish caught are tiny. A gold rush is on for the remaining bluefin tuna, and other large predators are declining at a vertiginous rate. It is a textbook fisheries management disaster.

Illegal fishing is rife in European waters because fishermen have little confidence in the system, enforcement falls to the twenty-five member states, and the penalties are low. Spanish fishermen catch 60 percent of their hake illegally and Scottish fishermen 50 percent of their cod. This compares with illegality of below 3 percent in Iceland and on the Pacific West Coast of America. In New Zealand if a skipper is convicted of underreporting catches or misreporting the position where he caught them, his catch is confiscated, his license withdrawn, and his vessel impounded. Separator trawls that catch haddock and release cod have been shown to work, but they play no part in EU policy.

The European Union's system of catch quotas and days at sea is so complicated it invites dishonesty. Regional fisheries councils, intended to emulate those in the United States, have no power. New quotas and days at sea are agreed by ministers from all the EU's twenty-five member states in one frantic three-day session in Brussels before Christmas each year, where ministers actually mess around with the law for their own political ends. Unsurprisingly, scientific advice is often ignored, particularly on the most overfished stocks. In these circumstances, plans to restore cod and other overfished stocks to their previous abundance have no credibility. Subsidies for building new vessels were phased out, but Mediterranean nations have successfully arranged for them to be reintroduced in exchange for more fuel-efficient engines.

Having ruined its own fisheries, Europe now exerts a malign influence across the globe. The demand of its 459 million people for fish is now comparable to the demands of Japan's 127 million, who have traditionally depended on fish for a much larger proportion of their diet. Its fishing industry is mired in illegality. As Sir Tipane O'Regan, former chairman of Sealord, the New Zealand fishing group that catches fish in New Zealand and Latin America, put it: "Most of the major pirate fleets are now owned by European capital—mostly Spanish or Russian—and on the whole industry knows who they are."

Europe's problems partly have to do with its confused constitutional structure. Though some countries such as France and Germany want the European Union to be a federal superstate, the EU in practice is a community of independent states with different jurisdictions. The fatal flaw in its fisheries policy is arguably fundamental, that it is common and that the common is undermanaged. Little notice is taken by Europe's fisheries managers of the demands of Europe's recreational fishermen or the environment.

What Ransom Myers said of the Canadian system is also true of Europe's: the system is set up to fail. The EU's management of its fisheries has many parallels with the Easter Islanders' management of their forests—which led their civilization ultimately to collapse—only the Easter Islanders did not have the opportunity to export their demand for timber.

By contrast, the United States has some successes to celebrate. On the positive side, there is a large, well-managed fishery in the Gulf of Alaska and the Bering Sea for pollock, large shrimp and recreational fisheries in the Gulf and on the East Coast, and successful rebuilding programs for Atlantic sea scallops, herring, black bass, and striped bass. On the negative side, the management of highly migratory species, such as swordfish and bluefin tuna, ranges from poor to bad. Gulf of Mexico and Atlantic reef fish are overfished. In New England, where overfishing has been going on longest, some ten out of nineteen stocks of groundfish remain overfished, but overall the biomass of this complex of groundfish species has rebounded by a third since 1996. The cod remains in an uncertain state, posing some of the most difficult questions for policy makers in any of the world's fisheries. Far greater cuts in fishing effort and far more precaution are warranted there.

To be brutally honest, America's improved handling of its fisheries is the direct result of doing the wrong thing for years. Overfishing by domestic and foreign fleets in the 1960s and 1970s had already severely depleted the Northeast's groundfish. The regional fisheries councils set up by the Magnuson Fishery Conservation and Management Act in 1976 failed to produce recovery, as they were dominated by fishermen's interests and the capacity of the domestic fleet proved excessive.

The nadir for the Georges Bank and Gulf of Maine came in

the early 1990s—but so, too, did a groundbreaking legal action by the Boston-based Conservation Law Foundation and the Massachusetts Audubon Society challenging the fishery managers to enforce the Magnuson Act and to halt overfishing of cod, haddock, and yellowtail flounder. As a result of this successful action, a plan was drawn up, Amendment 5, closing the fishery to new vessels, increasing the size of mesh, and reducing the number of days at sea. Even as Amendment 5 was coming into force, on March 1, 1994, scientists were reporting that Georges Bank haddock and southern New England yellowtail flounder stocks had collapsed and that the Georges Bank cod was in "imminent danger" of collapse. So the National Marine Fisheries Service closed 6,500 square miles of Georges Bank and southern New England waters to protect fish.

Despite their earlier success in court, conservationists remained convinced that the Magnuson Act had basic structural flaws because it did not require depleted stocks to be rebuilt as quickly as possible. It allowed regional management councils to set fishing levels above those that were biologically sustainable, and it was silent on bycatch and fish habitat. A network of a hundred organizations lobbied Congress to address these deficiencies.

The result was that in 1996, during an era when a Republican-controlled Congress was attempting to roll back environmental legislation across the board, a unanimous Congress enacted the Sustainable Fisheries Act. The Conservation Law Foundation has returned to the courts once more since then, testing whether the regional management council was actually doing what the 1996 act said it should. The court ruled the managers still weren't doing it right, and the fix to that was put in place in 2004.

There is new debate about the reauthorization by Congress of the Magnuson-Stevens Act (as it was renamed in 1996), now six years overdue. This is expected to bring in rights-based quotas for the first time and sector allocations, as they are called, whereby a cooperative gets the quota and decides whose boat will actually fish it and when. This will give rights and clarity and with luck help to further restrict fishing effort.

The New England fishery now has the largest mesh size in the world, at 6½ inches (compared with between 3 and 4 inches in Europe). As Frank Mirarchi, the Scituate fisherman, put it: "A fish has to be 20 inches long for it even to slow them down." There remains 8,500 square miles of fish habitat closed to fishing for habitat protection—compared with none in Europe. Fishing is managed by days at sea, trip limits, and closed areas. Days at sea give scientists headaches but are simple to enforce by sea and by air. There is observer coverage. There are opportunities for innovative fishing companies and cooperatives to fish in ways that minimize bycatch of overfished stocks and juveniles, typified by the Chatham hook fishery for cod and haddock. There is a feeling in the industry itself that science-based conservation works and that the pain has been worth it. Louis Linquata, the veteran Gloucester fish buyer, surprised me by saying, "You know, one day I think this will go over to a hook fishery again."

There is a can-do attitude to conservation in the New England fishery and much less whining about scientists being wrong. There are many well-educated people in the fishery. Chris Glass is a scientist who used to work at the University of Aberdeen in Scotland and is now director of the Northeast Consortium, a group of academic and research institutions, including Woods Hole, the University of New Hampshire, and the Massachusetts Institute of Technology, so he understands both sides of the Atlantic better than most. He told me: "We [in New England] have used many techniques that have been studied elsewhere, and for whatever reason were never adopted, because the industry [here] is receptive to using new approaches and new techniques to solve the problem." Paul Howard, former coast guard captain and executive director of the New England Fishery Management Council, emphasized the new cooperation between fishermen, scientists, and managers: "We're starting to get the same narratives. We've minimized the bullshitters. We talk about science, biology, economics, and trade-offs and marine mammals. Unless you have an all-inclusive dialogue, you are not going to get that narrative agreed to. I don't know if my friends in the UK [he has friends in the Scottish white fish industry] would hold a meeting with managers and scientists."

To my surprise, Howard gave credit to the environmental community for changing that common narrative. He said: "They have talked about [the fishing] as a public resource. It doesn't belong to fishermen. Fishermen know now that the fish have to feed the seals and the whales and the birds. It has been done through education and campaigning. Our Congress has listened and strengthened the conservation laws." What helps, above all, says Howard, is success—after years of closing fishing grounds, reducing effort, and driving everyone crazy. "We have recently started to see the curves go up."

You have to set against this optimism, of course, the continuing refusal of cod stocks to rebound. George's Bank and Gulf of Maine cod populations have plummeted 25 and 21 percent, respectively, since 2001. It is clear that New England will need all the legal expertise of conservationists and innovative conservation-oriented fishing methods it can muster if the cod is to recover. Fishermen are still fishing the cod at three times the sustainable rate, and there is only a tiny breeding stock left. There is more trouble on the horizon with the lobster, which is managed by the states, as it is mostly inside 3 miles. Striped bass eat lobster, and as they surge back, the lobster is tailing off. It started in Connecticut and is working its way north.

There are, of course, grumbles. There are complaints about the number of dogfish that fishermen are not allowed to land. There are noises about the big midwater trawlers, supposedly catching herring but which have been catching whales, porpoises, and groundfish, even anglerfish. That means their nets are on the bottom, said fisherman Tim MacDonald of Gloucester. "We're just eking out a living. They say the fish are coming back. Even if they do, there's still too much effort out there," was his view. And he's almost certainly right.

The New England example is both heartening and indicative of just how hard it is to recover fisheries when they collapse. Frank Mirarchi claims he is catching around 300 pounds an hour compared to 50 pounds a year or so ago. He told me: "I can't tell you how hard people have worked at this. It is an industry that has gone through complete cultural turmoil. It has taken fifteen years to show that fisheries management works. I'm not going to argue that nobody's cheating. But by and large people are in favor of all this. We're not going back to stealing fish. We've worked incredibly hard and diligently to get where we are. Only time will tell whether stocks will rebuild to what they were before Columbus got here."

His confidence is real. It has solid form. We were talking aboard the boat Frank and his thirty-year-old son Andrew recently built. They thought about it hard, then seized the moment. "It's the beginning of something," says Frank. "Whether it will work I don't know."

On land it has taken forty-five years, and a succession of food scares, since our first understanding of what intensive agriculture was doing to the land, for farmers to start talking about traceability, sustainability, and green farming and for organic food to become mainstream. I suspect it will take a comparable time to come to grips with the global crisis caused by intensive fishing. Yet I think we know what to do. A lot can be accomplished locally and regionally if federal and international bodies won't get their acts together. It is not rocket science. And we have a better chance of success in this enterprise than in beating global warming.

Looking back over my trips around the world's fisheries, I am struck by the fact that the happiest fishermen I met were those who owned, or effectively owned, their share of the resource and could exploit it when they wanted, whether it was the skipper of a huge Icelandic trawler catching cod very slowly, to maintain peak quality, or a lobster fisherman in a Newfoundland protected area marking the tail of a young lobster before putting it back and expecting others to do likewise. The only happy conservationists I met were those who controlled their part of the sea, on behalf of the public.

I came back believing that the time has come to change the laws of the sea so that they are more like the law of the land. It may seem paradoxical to suggest that the citizen is the true owner of the sea and that we should grant secure long-term rights to fish. With these rights, however, should come new responsibilities and a new citizenship ethic.

Technology has caused the crisis in the oceans. It can also be used to solve it by monitoring fishing boats wherever they happen to be. Why not put satellite monitoring data on the Internet and let the owners of the sea, us citizens, monitor the Bering Sea pollock boats or the *Atlantic Dawn* in real time? At the moment satellite data are regarded as confidential. The hunter-gatherer's knowledge of the fishing grounds is his intellectual property, his advantage over his competitors. Remove the race to fish by granting real property rights and there would be no need for confidentiality.

We need to fence the range, even in the wildest and remotest parts of the ocean. And we should not weep for the death of the cowboy.

20

RECLAIMING THE SEA

I pointed to the Surrey bank, where I noticed some light plank stages running down the foreshore, with windlasses at the landward end of them, and said, "What are they doing with those things here? If we were on the Tay, I should have said that they were for drawing the salmon nets; but here—"

"Well," said he, smiling, "of course that is what they are for. Where there are salmon, there are likely to be salmon nets, Tay or Thames; but of course they are not always in use; we don't want salmon every day of the season."

I was going to say, "But is this the Thames?" but held my peace in my wonder, and turned my bewildered eyes eastward to look at the bridge again, and thence to the shores of the London river; and surely there was enough to astonish me. . . . The soap works with their smoke-vomiting chimneys were gone; the engineer's works gone; the lead-works gone; and no sound of riveting and hammering came down the west wind from Thorneycroft's.

—William Morris, *News from Nowhere, an Utopia,* 1890

Lowestoft, September 2090. The herring is in. The sailing drifters, the only vessels allowed to fish on the edge of the town's marine reserve, have been filling their boxes. The scene is reminiscent of a seventeenth-century Dutch painting—only with modern clothes, navigation technology, and safety equipment. The inshore-caught herring, a great bounty that has opened up again since the reseeding of the Dogger Bank with herring spawn, is greatly prized, and the restored seafront market is full of tourists who want to buy some from the inshore fishermen who catch the "silver darlings" in traditional hemp nets.

Some tourists opt for the whole experience on offer and go out at night with the herring fishermen, who make a good living from the business that is now called heritage fishing. Others go to restaurants to have the herring fried in oatmeal by the chef, a Scottish tradition, and to eat oysters from areas carefully reseeded out at sea. Wild oysters were reseeded in the great experiment of restoring the Dogger Bank around 2040 and have been building up steadily, but the law says they may only be fished, like the inshore herring, from sailing vessels. Locally smoked kippers are in season and fetching good prices in the fish market, but the best prices of all go to the huge hand-lined sole caught on the edge of the no-take reserve. The *East Anglian Daily Times* newspaper says that there have been so many herring and anchovy off the east coast that bluefin tuna, which have not been recorded off the English coast since the 1950s, have been spotted in the reserve off Scarborough.

A few hundred yards along the coast, children in wetsuits have been pursuing a giant skate all morning, watched over by the great-grandson of one of the last Lowestoft trawlermen—a warden in the town's marine reserve. The skate has no fear of people and consents to be stroked before billowing away across the bottom. It can head away for miles in each direction quite safely, for bottom gear is banned in this section of the North Sea. The glass-bottomed boat has been following it, too. You can see well enough, though the visibility is only about 30 feet. The water has been clearing over the past five years thanks to the buildup of oysters on the bottom. The other great excitement on the edge of the reserve today was a pod of bottlenose dolphins, swimming among the wind farms out on the edge of the reserve and then playing tag with the returning drifters.

Captain Nemo's restaurant on the restored seafront is doing a roaring trade in fish from the inshore boats, including species such as red mullet, which would not have been thought of as North Sea fish a few years ago. English white wine, now grown as far north as Yorkshire, goes well with the fresh, salty fare. The younger clientele arrive in wetsuits after sailing or swimming. The menu is full of local fish—sole, catfish, haddock, and whiting—caught on regulation jigs and hand lines. An alternative supply comes from the haddock and witches (deepwater sole) caught by the twenty or so quota trawlers in the northern North Sea fishing grounds. The trawler fishermen accuse the inshore fishery of being a theme park, but it is actually providing something that people want and rich nations can afford—a way of being part of nature and the maritime heritage. It brings in tourists, creates jobs, and has given new life to the town. All the new building on the seafront is causing the city elders a headache in keeping up with the planning conditions, particularly since sea levels are a yard higher now and the amount of land protected by seawalls has shrunk. But the town has turned around in two ways since the early part of the century, economically and physically. All it once had was a much-reduced port and the Norfolk Broads wetland, to the back of the town, which have lost their luster since the creation of the Great Sole Reserve.

What children growing up today think of as entirely normal, educated adults know to be a great achievement. In the twentieth century and the early years of this century the world's oceans were in a sorry state. Europe in the twentieth century gradually dealt with the problems of pollution that had plagued the continent in the nineteenth century, killing off the salmon in rivers and making canals and waterways so poisonous that you had to have your stomach pumped out when you fell in them. But the problems of capture technology in the oceans had been underestimated and have taken nearly another century to tackle. Even then, the reforms have worked only in the advanced democracies and in the regional organizations that are based in them. There continue to be problems around nations with enormous populations, such as Indonesia, the Philippines, and, of course, along the coast of Africa, where the great upwelling fisheries of the west coast are now as much of a memory as its rain forests.

In the North Atlantic we finally cracked it, but it took the near extinction of the bluefin tuna in the eastern Atlantic and the Mediterranean, a real fright over the New England cod, and the commercial extinction of the cod in the North Sea—though a dispute goes on to this day about how much climate change was responsible. Forty years later we are still attempting to reseed the North Sea with cod from Norwegian farms. It is possible that the climate, or the ecosystem, may have changed too much for them ever to return to their former abundance. But the variety of fish in the sea has if anything increased. The present-day arrangements of inshore fishing licenses being limited to the least damaging methods and the offshore trawlers being run by a few companies with a financial interest in leaving fish in the sea date from the 2020s, as does the network of large marine reserves that began to restore the oceans' abundance and which was seized on by politicians as a popular measure and therefore grew.

It was almost certainly the Oceans Summit in 2012, twenty years after the Earth Summit in 1992, that set in place the laws needed to govern the middle of the oceans, but the reform of inshore waters in Europe and the creation of reserves actually has it origins in the period 2008–12, when a few reserves began to be created by law. There was a great dawning of public awareness about the oceans at the time of the Oceans Summit, which led to the oceans becoming the responsibility of the deputy prime minister in most European governments. Until then, the fishing industry had held sway over the sea.

Paradoxically, the few people who do fish now are much wealthier than they were then, and so are the shore-based processing industries and the tourism industries that have grown up around the new, strictly limited forms of fishing. The tracing of fish to the ocean, vessel, and method by which they were caught became popular in the first decade of this century and is taken for granted today. It was around that time that the premium for wild fish began and toppled some of the intensively farmed fish, such as salmon. Gradually many people developed a preference for some of the plentiful but overlooked fish that used just to be boiled up for meal. There's a restaurant on the front at Lowestoft, just past the statue of Michael Graham, that has specialized for years in recipes for sand eel and blue whiting.

Lowestoft, 2006. We can all dream. But when it comes to the oceans, we don't do it enough. The consequences of going on as we are hardly bear thinking about. In another fifty or a hun-

dred years all the world's oceans will be like the Mediterranean
or the Java Sea—largely empty of everything but tiny fish attempting to evade the nets long enough to reach breeding age
and mostly failing. The Mediterranean will be even more like a
giant built-up swimming pool because the swordfish and bluefin
tuna will have disappeared as surely as the bison did from the
great plains of North America. The last ones are bred using
technological methods that have been forced on the industry by
necessity. The great upwellings of Africa will be almost gone in
terms of their fisheries, with the exception of the southern hake
off Namibia, which will fetch huge prices in Spain.

Some people in countries with growing populations near the
coast will undoubtedly starve. The amount of fish off West
Africa will be so small that Senegal and other countries once
rich in fish will need food aid, just as Ethiopia and the Sahel do
now. Fish farming will rule and poison even the open sea. The
newspapers will be full of stories about diseases and mutations
that farmed fish cause to the remnant wild fish populations.
There will be more farmed fish than there is now, fed on soy-
beans and laced with genetically modified additives that mimic
fish oil, but these will taste quite different from wild fish, which
will fetch enormous prices.

To reverse these trends, we need to do several apparently
simple things:

- Fish less. If we fished at about half the pace or less that we
  do now, the bounty of the oceans would grow and we
  could eventually harvest more.
- Eat less fish, or eat fish less wastefully caught.
- Know more about what we are eating, and reject fish
  caught unsustainably.
- Favor the most selective, least wasteful fishing methods.

- Give fishermen tradeable rights to fish, accompanied by new responsibilities.
- Create reserves that will cover the migration hot spots for the big game fishes, such as tuna and swordfish, on the high seas and 50 percent of the entire area of intensively fished and used places, such as the North Sea.
- Make regional fisheries bodies responsible for the high seas work properly instead of merely monitoring the decline of the populations they are meant to preserve.
- Within our countries' 200-mile limits, we must organize a quiet democratic revolution, whereby the citizens regain overall control of the sea.

Wanting these things enough will bring them about. Already the market in the United States and Europe is switching toward fish managed in the right ways. As yet there are not enough of them. But the pressure is now on the fisheries that are the least sustainable. They are most likely to be rejected by the inquiring reader of a menu, or the concerned consumer who knows a thing or two at the fish counter or the freezer. The products of those fisheries, such as Georges Bank or North Sea cod, are likely to find their markets dwindling among the major retailers—generally far more sensitive to customer preference and environmental campaigning than politicians. The fisheries of the future are going to have to come up with reasons why consumers should think they are sustainable.

Remember what John Gruver said: "The fisherman of the future isn't going to be measured by the fish he does catch, but by the fish he doesn't catch." To begin measuring what fishermen do, we need to know what fish we are being offered to eat—what kind of tuna it is as well as how and where it was caught—and if we aren't told, we should decline loudly, so others can overhear. An awful lot of work will have to be done in the fisheries of the world to produce these fish that we want to buy. The problem at the moment is that the examples of best practice in the ocean lie dotted around the world in little puddles, like the fast-dwindling cod on Georges Bank and in the North Sea.

We have on offer two futures. One requires difficult, active choices starting now. If we don't take those choices, the other future will happen anyway.

CHOOSING FISH: A GUIDE

This book is not a consumer guide. At least that was not my primary intention. This has been more of a personal journey through the world's oceans, looking at fishing methods and their effect on target species and the rest of the marine ecosystem. On the way, though, I have come to realize just how much consumer choice can make a difference in changing the way that oceans are managed and fishermen pursue their catch. Therefore, I think, it would be a failure of nerve not to include the briefest conclusions about what to avoid eating, what to be cautious about, and what to eat without a twinge of conscience.

The list is necessarily idiosyncratic. It defers to, but sometimes disagrees with, consumer guides produced by the Monterey Bay Aquarium, the National Audubon Society, the Blue Ocean Institute and, in Britain, the Marine Conservation Society. These guides, which are on the web, are, rightly, the first port of call for anyone trying to decide the ethics of what to eat because they are updated regularly and have input from a wide range of scientific opinion. I would recommend eating anything certified or in the process of being certified by the Marine Stewardship Council (i.e., that has passed through the precertification process). I am less convinced by other certification schemes, none of which are as global or involve so many shades of opinion in the certification process. I recommend FishBase and the IUCN Red List, both on the Web, as a means of resolving after-dinner arguments about fish names, which fish are endangered, and which are fast or slow growing, vulnerable, or less vulnerable to overexploitation.

Ultimately, consumers need enough information about fish to make their own decisions about what to eat. So lack of information is as good a reason for not buying something as any other. This means that I consider it unacceptable to sell something called "tuna" without saying which species it happens to be and where and how it was caught. As a matter of personal choice, I now refuse to eat any fish that is not identified as to its species, ocean of origin, and method of capture—which also means turning down things called "Tuna crunch sandwich filler" and "Italian tuna steaks" (was the tuna Italian or the way of preparing them?) By rejecting fish that is not properly labeled consumers vote with their dollars for a world better informed about what is going on in the sea.

So here, with profuse acknowledgments to all the above sources of information, is what I would personally rather not eat—unless the management of stocks improves radically—and what I would choose if it were on the menu.

FISH TO AVOID

Atlantic cod Most are from spawning populations recognized as overfished, particularly Newfoundland, Georges Bank, Gulf of Maine, North Sea, and Norway/Russia. Icelandic cod are okay, with the caveats made earlier in the book. Cod caught on rod and line would be acceptable, as would farmed organic cod fed only on meal made from offcuts of other fish.

Atlantic haddock Haddock are usually caught with cod, so although they are currently plentiful on both sides of the Atlantic they should be avoided unless the seller can prove that they are caught in selective trawls that actively exclude cod. It can be done, which makes you wonder why it is not always done.

Atlantic halibut Listed as endangered by IUCN. Slow growing and needs a break. If you must, choose line-caught from Icelandic waters.

Bluefin tuna Disgracefully overfished on both sides of the Atlantic and in the Mediterranean. As one of the oceans' most astonishing and endangered mega-fauna, it is long overdue for a listing under the Convention on International Trade in Endangered Species. Environmentalists, here is something worth campaigning about.

Caviar Given the price of the stuff, and the fact that there is a ban on international trade in most forms of caviar as this book goes to press, most of us can feel doubly virtuous avoiding wild-caught caviar. Farmed American or French caviar provides a viable alternative.

Chilean sea bass (Patagonian toothfish) Avoid, except MSC certified fish from around South Georgia and traceable fish from now better-policed Australian waters around Antarctica.

Grouper Most species of these slow-growing reef fish are overfished. Consult FishBase.

Orange roughy As with round-nosed grenadier, these are almost certainly unsustainably caught, with the just-possible exception of roughy from New Zealand. Maintaining catches means damaging more virgin sea mounts.

Sharks, skates, and rays Sharks of all species are slow to mature, have few young, and are vulnerable to overfishing. Sharks are caught not only for their meat but also for their fins. Fins, favored by the Asian shark-fin soup market, are hacked off and the sharks are thrown back into the sea to die, a barbaric practice. Skates and rays are also slow to mature and some are threatened with extinction according to IUCN.

Snapper Avoid red snapper in the United States, which is overfished.

Swordfish Slow to mature top predator that plays important role in the ecosystem. North Atlantic stock is assessed as vulnerable by IUCN. As with sharks and some tunas, consumption should be limited because of concern about mercury.

FISH TO BE WARY ABOUT

Shrimp The bycatch, generally huge, is the big issue with the wild ones. Trap-caught or responsibly farmed ones are the ethical choice. U.S. farmed or wild caught are acceptable. Avoid imported.

Tuna Purse-seined tuna, used in cans, is worrying for several reasons. It may contain endangered bigeye, listed as vulnerable by IUCN, and involves an awesome, but often unquantified, bycatch. We need more information from the tuna canners about how much bycatch there is and how tuna was caught. There is no reason to feel any happier because your can is marked "dolphin friendly."

Long-lining for tuna may have an even worse bycatch of endangered and slow-growing species than purse-seining. Then again it may not—the evaluations do not exist. One thing is for sure, it kills a heck of a lot of sharks and turtles. Troll or pole and line-caught skipjack, albacore, and yellowfin are the best alternative.

FISH TO EAT WITH A CLEAR CONSCIENCE

Blue whiting Small cod-like fish that could substitute perfectly well for cod in the countries that are fond of it, but in the process of being massively over-exploited for fishmeal. Building a market for it as a table fish, which would mean it was fished for more slowly instead of being caught and pulped for fishmeal, could help to keep it from collapse.

Herring Fast reproducing, oily fish. Reasonably well managed on the East Coast of United States and in Europe. One of the best sources of no-regrets omega 3s.

Hoki MSC certified and apparently within biological limits, though catches have been reduced.

Horse mackerel Oily fish beloved of Japanese emperors. Usually finds its way into fish meal. Why not eat it instead of farmed salmon?

Lobster Some stocks are overfished, but potting is selective and there are plans to "ranch" lobster to preserve stocks.

Mussels and oysters These are farmed in ways that have minimal impact on the environment. Scallops can be, too, but the wild ones are often caught by dredging, which damages the bottom.

Pacific halibut Certified by the MSC as well managed.

Pacific salmon Alaskan salmon is far better conserved than the Atlantic kind. It is also MSC certified.

Pollock Certified by MSC. There are some bycatch issues, both of sea lions and salmon, but these are relatively well managed. Overfished in Russian Pacific waters.

Sand eel/sand lance This was the staple of European fishmeal plants, until its population crashed. Perfectly acceptable fried in batter. So why don't people eat it instead of feeding it to salmon at a 3:1 conversion rate? Same goes for capelin.

Sardine Pacific sardines once supported the largest and most profitable U.S. fishery, which nearly disappeared in the 1940s. Overfishing is thought to have helped to precipitate this decline, but the population was known to have a "boom and bust" cycle. Sardines—small members of the herring family—have made a comeback on both coasts. They reproduce fast and have plenty of omega 3s.

Striped bass Farmed and wild caught.

Tilapia Vegetarian fish are the one true hope of the developing world and could be the future of ethical fish farming in developed countries too. Might as well get used to cooking them now. Over to you, celebrity chefs.

ACKNOWLEDGMENTS

At times I still find it difficult to believe that *The End of the Line* actually got written, seven years and two agents after I tried to interest publishers in such an idea, and thirteen years after it first seemed to me that the world needed such a book. It has taken another two years to get published in the United States, albeit in a different version to the one that was published in Europe, the Commonwealth, and Japan. I am now reconciled to the fact that this book is one of those publishing stories about trying and trying and only eventually succeeding.

In 1997, my then agent, Xandra Bingley, who has gone on to take up writing full time, was kind enough to go the rounds of London publishers with a proposal. We found that publishers thought the subject of what commercial fishing was doing to the world we live in, and the fish we eat, interesting but not a commercial proposition. Another agent tried three years later. Then, in 2003, just when I had given up hope of ever interesting anyone in the idea, I got a call from Brendan May, a friend and chief executive of the Marine Stewardship Council, to say that he had been approached by an agent with who wanted to find someone to write a book about fish. His name was Ivan Mulcahy. Brendan told Ivan I had probably already written it, which was not true, but it meant Ivan made the call that began the story of this book.

So my thanks must begin with Brendan—whose organization I have therefore felt obliged to give as thorough a turning-over as I have anyone else's—for his faith in me and for introducing me to Ivan, who persuaded me to write this book, made it possible for me to take the time to do it, and who was responsible for recognizing my better ideas, deleting my worst ones, and for reading every word several times over in the hope that I might live up to my original intentions. I owe both a great debt. After them, I must thank Fiona MacIntyre at Ebury Press for taking it on with such confidence in the UK and seeing off what, to my surprise, was, by 2003, a great deal more interest in the subject of fish in the publishing business than had existed seven years previously. Thanks after that must go to those who wrote to publishers on my behalf to persuade them that this proposal was a project of some merit: I am grateful on this count to HRH the Prince of Wales, Margaret Atwood, Sebastian Faulks, Brendan May, and George Monbiot. I must also thank Charles Moore, then editor of the *Daily Telegraph*, where I work, for giving me the time off to write the book; the news desk for putting up with me at difficult moments; and the *Daily Telegraph* generally for sending me on one or two assignments which have ended up being part of it.

Perhaps it was the urgent need to tell the story of what is going on in the sea in the pursuit of what we eat that made so many people so very kind to me. Along the way, recently or not so recently, the following people offered invaluable help, either by being particularly generous with their time and ideas, or being kind enough to read and offer suggestions on my drafts,

25

or both. These were: John Beddington, Brendan May, Euan Dunn, Orri Vigfússon, Ragnar Árnason, Daniel Pauly, Joe Horwood, Mark Tasker, Jonathan Peacey, Sidney Holt, Monica Verbeek, Sue Windebank, Mike Sutton, John Gummer, John Shepherd, Tommy Finn, John Williams, John Pope, Raul Garcia, Georgina Mace, Papa Samba Diouf, Odd Nakken, Alan Simcock, Han Lindeboom, Bill Ballantine, George Clement, Alastair O'Rielly, Richard Sandbrook, Ken Stump, Michael Earle, Hélène Bours, Geoff Kirkwood, and my sister Gillian Clover.

I must thank those who gave up their time and contributed thoughts and encouragement, though often at some remove, and in one or two cases without ever having met me face to face—a new definition of kindness in the information age. These include Jake Rice, Ransom Myers, Neil Fletcher, Sir Tipene O'Regan, Rögnvaldur Hannesson, Chris Hutchings, Jenny Hodder, Jim Gilmore, Brian O'Riordan, and Thelma Wilson.

Then there were those who helped me when I was researching and writing the U.S. version of this book, an undertaking which saw new chapters 1 and 19, a new recreational fishing chapter, and a new version of the chefs chapter to address American, rather than European, chefs. I am particularly indebted to Chris Glass, Teri Frady, Priscilla Brooks, Ellen Pikitch, Andy Rosenberg, Frank Mirarchi, Pat Kurkul, John Boreman, Fred Serchuk, Shannon Crownover, and Paul Howard. The U.S. edition would never have happened without Colin Robinson at The New Press, who took it on, and my unfailingly rigorous but cheerful editor, Lizzie Seidlin-Bernstein, who demonstrated what I seem always to have known: that New Yorkers edit better than anyone else on Earth.

I must also thank all the people I interviewed for the book. As this is a nonfiction book most of their names are already recorded in it and, to save the reader's patience, I intend not to list them again; but everyone who is mentioned in the book will know how generously, without exception, they gave up their time in the hope that I would emerge better informed. I salute them for expressing their views so memorably, in ways that have become the very fabric of what I have written. If there is anyone who feels they should have been included, and they are not, mine is the journalists' excuse: this book was eventually written very fast, it was late, time pressed, and my brain was not at its best after writing from morning to late at night. My hope is that, when it comes to reporting what they actually said, or their ideas, I did not let them down. The last and most important thank-you is to my wife Pamela for putting up with me, supporting me when no one else did, and for reading every word that I wrote to make sure that, even if I was making a fool of myself, at least it was deliberate. My love and thanks to her.

—Charles Clover,
Dedham, England, 2006

GLOSSARY

CCAMLR Commission for the Conservation of Antarctic Marine Living Resources. Pronounced "kammelar." Scientific body set up in 1982 under the Antarctic Treaty to conserve marine life in the Southern Ocean. Takes particular account of the links between species, such as birds, seals, and krill—"the ecosystem approach." Enforcement leaves a lot to be desired.

CEFAS Centre for Environment, Fisheries and Aquaculture Science, the UK government fisheries science agency.

CITES Convention on International Trade in Endangered Species of Wild Flora and Fauna. Intended to ensure trade in wild animals and plants does not threaten their survival. Entered into force in 1975. List of protected species includes relatively few fish.

COLTO Coalition of Legal Toothfish Operators. Australia-based coalition that names and shames toothfish poachers in its Web-based Rogues Gallery.

Demersal fish Fish living and feeding on the bottom, such as cod, haddock, or hake. Sometimes also called groundfish.

DFO The Canadian Department of Fisheries and Oceans, inadvertent architects of the Grand Banks cod collapse.

EEZ Exclusive Economic Zone. The EEZ was a key provision of the 1982 UN Convention on the Law of the Sea (UNCLOS), allowing each coastal state exclusive rights over all resources, whether oil and gas, fish, gravel, or minerals within 200 nautical miles of its coast.

FAD Fish Aggregation Device. Fishermen noticed that tropical tuna congregated under logs, whales, and other large objects. Now they make their own: a FAD is a wooden frame with (optional) wisps of netting hanging from it. This is launched into the ocean currents attached to a radio or satellite beacon that can relay all sorts of information that indicates the likely presence of fish, including sea temperature, back to the fishing vessel.

FAO UN Food and Agriculture Organization. Pro-development UN organization that publishes reports on the state of world fisheries.

GPS Global Positioning System, satellite navigation system originally developed and funded by the US military, enabling users to compute position to within a few meters anywhere in the world.

IATTC Inter-American Tropical Tuna Commission. Established in 1950. Regional fisheries organization for the eastern Pacific, based in La Jolla, California.

ICCAT International Commission for the Conservation of Atlantic Tunas. Established in 1969 and based in Madrid. Known by conservationists as International Conspiracy to Capture All Tunas because of its indifferent success in conserving fish populations. Some successes in forcing pirate vessels off rogue states' registers and reducing swordfish catches in the North Atlantic.

ICES International Council for the Exploration of the Sea. Founded in 1902. The organization that promotes and coordinates marine research in the North Atlantic. More than one hundred years on, it still hasn't calculated the preindustrial spawning stock of the North Sea cod.

IOTC Indian Ocean Tuna Commission. Established in 1993. Regional fisheries body for tuna in the Indian Ocean, based in the Seychelles. (Nice work if you can get it.) Does stock assessments and compiles list of authorized fishing vessels but as yet does not apply any active conservation measures.

ITQ Individual Transferable Quota. Also known as Individual Fishing Quotas. Long-term tradable rights to fish which adherents claim have been highly successful in ending the "race to fish."

IUCN International Union for the Conservation of Nature and Natural Resources. Global conservation science network, also called the World Conservation Union. Publishes Red List of the world's endangered species on the Web.

IUU Illegal, unregulated, and unreported. UN-speak for pirate fishing.

IWC International Whaling Commission. Club of whaling nations established in 1946 to manage whale stocks. If fisheries were run by its latest management rules—developed since commercial whaling ended—hardly a cod boat would put to sea.

MSC Marine Stewardship Council. International nonprofit organization, based in London, which promotes sustainable fisheries through certification. Has the most demanding criteria and widest range of participants involved in the assessment of fisheries of certification organizations.

MSY Maximum Sustainable (or Sustained) Yield. Optimum relationship between fish stock and fishing effort, the pursuit of which has almost always led to the stock being overfished.

NAFO Northwest Atlantic Fisheries Organization. Established in 1979 and based in Dartmouth, Nova Scotia. Regional organization for waters beyond the continental shelf. There are sixteen members.

NEAFC North East Atlantic Fisheries Commission. Regional fisheries organization. Formally established in 1963 but dates back to between the wars. Based in London. Latest failure: blue whiting.

NOAA National Oceanic and Atmospheric Administration, an agency of the U.S. Department of Commerce and parent organization of the National Marine Fisheries Service.

OECD Organization for Economic Cooperation and Development, an economic research body funded by the top thirty industrialized nations.

PCBs Polychlorinated biphenyls—highly persistent fire-retardant chemicals used as fluid in transformers and capacitors from the 1930s. Manufacture ceased in 1977 in the U.S. after the discovery that their toxic effects built up in the organs of marine creatures as a result of dumping at sea.

Pelagic fish Fish living in the upper waters of the open sea such as mackerel, sardine, or tuna.

UNCLOS UN Convention on the Law of the Sea. Called for in 1967, talked about in 1973, adopted in 1982, but still not ratified by United States.

Whitefish Fish with white flesh whose main reserves of fat are in the liver, such as cod, haddock, or whiting. As opposed to oily fish such as herring, sprat, mackerel, or shellfish.

WWF In America, this refers to the World Wildlife Fund. Everywhere else it is the organization formerly known as the World Wide Fund for Nature (now annoyingly just known by its initials).

BIBLIOGRAPHY

These books provide a good introduction for those coming to the subject of overfishing for the first time. The list of further reading offers more suggestions for the general reader.

Chapter 1

Pauly, Daniel and Reg Watson. "Systematic Distortions in World Fisheries Catch Trends," *Nature* 414 (2001): 534–6.

Rosenberg, Andrew A., W. Jeffrey Bolster, Karen E. Alexander, William B. Leavenworth, Andrew B. Cooper, and Matthew G. McKenzie. "The History of Ocean Resources: Modeling Cod Biomass Using Historical Records." *Frontiers in Ecology and Environment* 3, no. 2 (2005).

Smith, Tim D. *Scaling Fisheries: The Science of Measuring the Effects of Fishing, 1855–1955* (Cambridge: Cambridge University Press, 1994).

Chapter 2

McGoodwin, James R. *Crisis in the World's Fisheries: People, Problems, and Policies* (Palo Alto: Stanford University Press, 1991).

Myers, Ransom and Boris Worm. "Rapid Worldwide Depletion of Predatory Fish Communities," *Nature* 423 (2003): 280–3.

Pauly, Daniel, Villy Christensen, Sylvie Guénette, Tony J. Pitcher, U. Rashid Sumaila, Carl J. Walters, R. Watson, and Dirk Zeller. "Toward Sustainability in World Fisheries," *Nature* 418 (2002): 689–95.

Pauly, Daniel. "Why the International Community Needs to Help Create Marine Reserves." Fourth meeting of the United Nations Open-ended Informal Consultative Process on Oceans and the Law of the Sea, New York, 4 June 2003.

Safina, Carl. *Song for the Blue Ocean: Encounters Along the World's Coasts and Beneath the Seas* (New York: Henry Holt, 1998).

Chapter 3

Environmental Justice Foundation. "Squandering the Seas: How Shrimp Trawling is Threatening Ecological Integrity and Food Security around the World" (www.ejfoundation.org, 2003).

Institute for European Environmental Policy. "Fisheries Agreements with Third Countries—Is the EU Moving Towards Sustainable Development?" (www.ieep.org.uk, 2002).

Chapter 4

de Groot, S. J. and H. J. Lindeboom, eds. "Environmental Impact of Bottom Gears on Benthic Fauna in Relation to Natural Resources Management and Protection of the North Sea" (Netherlands Institute for Sea Research, 1994).

International Council for the Exploration of the Sea. "Environmental Status of the European Seas" (www.ices.dk, 2003).

Lindquist, Ole. "The North Atlantic Grey Whale *(Escherichtius robustus):* An Historical Account Based on Danish-Icelandic, English and Swedish Sources Dating from *c.* 1000 to 1792," Centre for Environmental History and Policy, Occasional Papers 1, Universities of St Andrews and Stirling (2000).

Naylor, Paul. *Great British Marine Animals* (Sound Diving Publications, 2003).

Nichols, J., T. Huntington, P. Winterbottom, and A. Houg. "Certification report for the Pelagic Freezer Trawler Association North Sea Herring Fishery" (www.msc.org, 2003).

Olsen, O.T. *The Piscatorial Atlas of the North Sea, English and St George's Channels* (London: Taylor & Francis, 1883).

Pauly, Daniel and Jay MacLean. *In a Perfect Ocean: The State of Fisheries and Ecosystems in the North Atlantic Ocean* (Washington, D.C.: Island Press, 2002).

Wigan, Michael. *The Last of the Hunter Gatherers: Fisheries Crisis at Sea* (Shrewsbury: Swan Hill Press, 1999).

Chapter 6

Aikman, Phil. "Is Deep Water a Dead End? A Policy Review of the Gold Rush for 'Ancient Deepwater' Fish in the Atlantic Frontier" (Greenpeace, 1997).

Gordon, John D. M. "Fish in Deep Water," Scottish Association for Marine Science newsletter 27 (2003).

Gordon, John D. M. "Rockall Plateau now in International Waters. The Fleets Move In," Scottish Association for Marine Science newsletter 22 (2000).

Gordon, John D. M. "The Rockall Trough, Northeast Atlantic: An Account of the Change From One of the Best-Studied Deep-Water Ecosystems to One that is Being Subjected to Unsustainable Fishing Activity," North Atlantic Fisheries Organization, Scientific Council Report N4489 (2001).

Lack, M., K. Short, and A. Willock. "Managing Risk and Uncertainty in Deep-Sea Fisheries: Lessons from Orange Roughy" (TRAFFIC Oceania and WWF Australia, 2003).

Chapter 7

Beverton, Ray, and Sidney Holt. *On the Dynamics of Exploited Fish Populations* (London: Ministry of Agriculture, Fisheries and Food, 1957).

Dyson, John. *Business in Great Waters: The Story of British Fishermen* (London: Angus & Robertson, 1977).

Graham, Michael. *The Fish Gate* (London: Faber & Faber, 1943).

Finlayson, Alan Christopher. *Fishing for Truth: A Sociological Analysis of Northern Cod Stock Assessments from 1977 to 1990* (Newfoundland: Institute of Social and Economic Research, Memorial University, 1994).

Huxley, Thomas Henry. Address to the International Fisheries Exhibition, London, 1883.

Lee, A. J. *The Directorate of Fisheries Research: Its Origins and Development* (London: Ministry of Agriculture, Fisheries and Food, 1992).

MacKenzie, Debora. "The Cod that Disappeared," *New Scientist* (16 September 1995).

Rozwadowski, Helen M. *The Sea Knows No Boundaries: A Century of Marine Science Under ICES* (Seattle: University of Washington Press, 2002).

Walters, Carl and Jean-Jacques Maguire, "Lessons for Stock Assessment from the Northern Cod Collapse," *Fish Biology and Fisheries* 6 (1996): 125–37.

Chapter 8

Porter, Gareth. "Estimating Overcapacity in the Global Fishing Fleet" (WWF, 1998).

Rice, Jake C., Peter A. Shelton, Denis Rivard, Ghislain A. Chouinard, and Alain Fréchet. "Recovering Canadian Atlantic Cod Stocks: The Shape of Things to Come," paper given at a conference organized by International Council for the Exploration of the Sea: The Scope and Effectiveness of Stock Recovery Plans in Fishery Management, Tallinn, Estonia, 2003.

WWF/Spain. "WTO and Fishing Subsidies: Spanish Case." Summary of Spanish report in English. (WWF, 2003).

Chapter 9

Bours, Hélène (Greenpeace International). "The Tragedy of Pirate Fishing," International Conference against Illegal, Unreported and Unregulated Fishing, Santiago de Compostela, Spain, 25–26 November 2002.

Doulman, David J. "Global Overview of Illegal, Unreported and Unregulated Fishing and Its Impacts on National and Regional Efforts to Sustainably Manage Fisheries: The Rationale for the Conclusion of the 2001 FAO International Plan of Action to Prevent, Deter and Eliminate Illegal, Unreported and Unregulated Fishing" (FAO, August 2003).

Earle, Michael (Fisheries Adviser, Green/EFA Group in the European Parliament). "The European Union, Subsidies and Fleet Capacity, 1983–2002," UNEP Workshop on the Impacts of Trade-Related Policies on Fisheries and Measures Required for Sustainable Development, 15 March 2002.

Hardin, Garrett. "The Tragedy of the Commons," *Science* 162 (December 1968): 1243–8 and *The Concise Encyclopedia of Economics* (Library of Economics and Liberty, www.econolib.org, 2002).

Justinian, *The Institutes of Justinian*, Book Two.

Lack, M. and G. Sant, "Patagonian toothfish: are conservation and trade measures working?" *TRAFFIC*, issue 1, vol. 19 (2001).

Chapter 10

Suarez de Vivero, Juan L. and Juan C. Rodriguez Mateos. "Spain and the Sea: The Decline of an Ideology, Crisis in the Maritime Sector and the Challenges of Globalisation," *Marine Policy* 26 (2002).

Chapter 11

Clarke, Bernadette. *Good Fish Guide* (Marine Conservation Society, 2002)

Environmental Justice Foundation. "Squandering the Seas: How Shrimp Trawling is Threatening Ecological Integrity and Food Security around the World" (www.ejfoundation.org, 2003).

Matsuhisa, Nobuyuki. *Nobu: The Cookbook* (London: Quadrille, 2001).

Seafood Watch Program (see Monterey Bay Aquarium in Websites).

Chapter 12

Joseph, James. "Managing Fishing Capacity of the World Tuna Fleet," *Fisheries Circular* 982 (FAO, 2003).

Mina, X., I. Artetxe and H. Arrizabalaga. "Updated Analysis of Observers' Data Available from the 1998–99 Moratorium in the Indian Ocean," IOTC proceedings 5 (2002): 340–5.

Romanov, E. V. "By-catch in the Purse-seine Tuna Fisheries in the Western Indian Ocean," Seventh Expert Consultation on Indian Ocean Tunas, Victoria, Seychelles, 9–14 November 1998.

Suzuki, Ziro (National Institute of Far Seas Fisheries). "Memorandum on Regulatory Measures for Purse-seine Fisheries in the Western Tropical Indian Ocean" IOTC proceedings 5 (2002): 176–7.

Chapter 13

Cunningham, Stephen and Jean-Jacques Maguire. "Factors of Unsustainability and Overexploitation in Fisheries" (FAO, February 2002).

Report and documentation of the International Workshop on Factors of Unsustainability and Overexploitation in Fisheries, Bangkok, Thailand, 4–8 February 2002 (FAO, 2002).

Rice, Jake. "Sustainable Uses of the Ocean's Living Resources," *Isuma*, Canadian Journal of Policy Research (Autumn 2002).

Chapter 14

Jones, Laura and Michael Walker, eds. *Fish or Cut Bait: The Case for Individual Transferable Quotas in the Salmon Fishery of British Columbia* (Fraser Institute, 1997).

Leal, Donald R. *Fencing the Fishery: A Primer on Ending the Race for Fish* (Center for Free Market Environmentalism, www.perc.org, 2002).

Chapter 15

Ballantine, Bill. "Marine Reserves in New Zealand: The Development of the Concept and the Principles," proceedings of an International Workshop on Marine Conservation for the New Millennium, Korean Ocean Research and Development Institute, Cheju Island, November 1999.

Ballantine, Bill. " 'No Take.' Marine Reserve Networks Support Fisheries," paper given at 2nd World Fisheries Congress, Brisbane, 1996 (These and other Ballantine papers can be found on www.marine-reserves.org.nz or www.auckland.ac.nz).

Gell, Fiona R. and Callum M. Roberts. "The Fishery Effects of Marine Reserves and Fishery Closures," World Wildlife Fund, Washington, 2003.

Chapter 17

Blue Ocean Institute, Seafood Guide (see Web sites)

Gilmore, Jim. "The MSC and Its Significance," At-Sea Processors Association, October 2003.

Guardian poll about the British public's ethical purchasing habits published 28 February 2004.

Marine Stewardship Council papers on pollock, hoki, Thames herring, etc., including those for and against certification, can be found at www.msc.org.

Monterey Bay Aquarium, Seafood Watch (see Web sites).

Wigan, Michael. "Breaking Covert: Salmon Farming," *The Field* (February 2006).

Chapter 18

Barlow, Stuart M. "The World Market Overview of Fish Meal and Fish Oil," Second Seafood and By-products Conference, Alaska (www.iffo.org.uk, November 2002).

Clover, Charles. "The Price of Fish," *Telegraph* Magazine (26 October 1991).

Hites, Ronald A., Jeffrey A. Foran, David O. Carpenter, M. Coreen Hamilton, Barbara A. Knuth, and Steven J. Schwager. "Global Assessment of Organic Contaminants in Farmed Salmon." *Science* 303 (2004): 226.

Humphrys, John. *The Great Food Gamble* (Abingdon: Hodder and Stoughton, 2001).

Information on deep-water fisheries and the biological status of the stocks is contained in the Advisory Committee on Fishery Management (ACFM) report (ICES, 12 May 2003).

"The Promise of a Blue Revolution," *The Economist* (7 August 2003).

Chapter 19

National Audit Office. "Fisheries Enforcement in England," a report by the Comptroller and Auditor General (April 2003).

Shoard, Marion. *Theft of the Countryside* (London: MT Smith, 1980).

FURTHER READING

Sylvia A. Earle, *Sea Change: A Message of the Oceans* (New York: Ballantine, 1996).

Richard Ellis, *The Empty Ocean* (Washington, D.C.: Shearwater, 2004).

Fishing News, a weekly newspaper based in the UK and aimed at fishermen.

Fishing News International, a monthly newspaper for fishermen.

Mike Holden, *The Common Fisheries Policy: Origin, Evaluation and Future,* with an update by David Garrod (Fishing News Books, 1994).

Mark Kurlansky, *Cod: A Biography of the Fish that Changed the World* (New York: Penguin, 1998).

Arthur McEvoy, *The Fisherman's Problem Ecology and Law in the California Fisheries, 1895–1980* (Cambridge, 1986).

Farley Mowat, *Sea of Slaughter: A Chronicle of the Destruction of Animal Life in the North Atlantic* (Boston: Houghton Mifflin, 1996).

Daniel Pauly, *On the Sex of Fish and the Gender of Scientists: A Collection of Essays in Fisheries Science* (New York: Springer, 1994).

Carl Safina, *Song for the Blue Ocean: Encounters Along the World's Coasts and Beneath the Seas* (New York: Owl Books, 1999).

WEB SITES

Blue Ocean Institute, Seafood Guide
www.blueoceaninstitute.org

FishBase lists 28,500 species of fish, with information about everything from their rarity and ability to withstand fishing pressure, to how they reproduce. www.fishbase.org

Food and Agriculture Organization of the United Nations
www.fao.org

Marine Conservation Society (UK), Good Fish Guide
www.fishonline.org

Monterey Bay Aquarium, Seafood Watch guide
www.mbayaq.org

National Audubon Society, Seafood Wallet cards
www.audubon.org

Red List of threatened species produced by the International Union for the Conservation of Nature and Natural Resources
www.redlist.org

Rogues' gallery of IUU fishermen
www.colto.org

Stock assessments in the North Atlantic compiled by the International Council for the Exploration of the Sea
www.ices.dk

INDEX

Check out these other books from The New Press

Blue Gold: The Fight to Stop the Corporate Theft of the World's Water
Maude Barlow and Tony Clarke
PB, $18.95, 304 pages
ISBN 978–1–56584–813–9
Nature/Environment & Ecology
The internationally acclaimed story of the corporate takeover of our most
basic resource and the inevitable global water crisis

*The Water Atlas: A Unique Visual Analysis of the World's Most Critical
Resource*
Robin Clarke and Jannet King
PB, $24.95, 128 pages
ISBN 978–1–56584–907–5
Nature/Environment
A comprehensive charting of the global water industry

Diet for a Dead Planet: Big Business and the Coming Food Crisis
Christopher D. Cook
PB, $17.95, 352 pages
ISBN 978–1–59558–084–9
Environment
A timely indictment of industrial agriculture's threat to the future of food,
health, and the environment

Gone Tomorrow: The Hidden Life of Garbage
Heather Rogers
PB, $15.95, 304 pages
ISBN 978–1–59558–120–4
Nature/Environment & Ecology
A sobering exploration of our high-octane trash output that was
named an editor's choice by the *New York Times* and a nonfiction
choice by *The Guardian*

The Body Hunters: Testing New Drugs on the World's Poorest Patients
Sonia Shah
HC, $24.95, 256 pages
ISBN 978–1–56584–912–9
Medicine
An eye-opening look at big pharma's unethical and exploitative drug trials in the global South

The Conquest of Bread: 150 Years of Agribusiness in California
Richard A. Walker
HC, $27.95, 400 pages
ISBN 978–1–56584–877–1
Nature/Agriculture
A sweeping analysis of California's agrarian history from 1850 to the present

Fallout: The Environmental Consequences of the World Trade Center Collapse
Juan Gonzalez
PB, $13.95, 160 pages
ISBN 978–1–56584–845–0
Current Affairs/Politics/Environment
What public officials failed to say about post–9/11 New York air

Visit us online at www.thenewpress.com